ENGINEERING CONCEPTS OF ELECTRICITY AND MAGNETISM

(FOR DIPLOMA AND DEGREE COURSES IN ELECTRICAL ENGINEERING)

notionpress
.com

ENGINEERING CONCEPTS OF ELECTRICITY AND MAGNETISM

(FOR DIPLOMA AND DEGREE COURSES
IN ELECTRICAL ENGINEERING)

UTPAL BASU

notionpress.com

Notion Press

Old No. 38, New No. 6
McNichols Road, Chetpet
Chennai - 600 031

First Published by Notion Press 2016

ISBN 978-93-86073-34-1

Dedicated with love and affection
to my wife Gopa and our daughter Upasana

Contents

Contents (P-viii)

Contents (P-ix)

Contents (P-x)

Foreword

Electricity and magnetism are two very important chapters in the basic science of physics, which are transformed into action or application in electrical engineering. Electricity concerns the properties and movement of electrically charged particles. Magnetism concerns the properties of magnets and magnetic materials. Electricity and magnetism are interrelated. Movements of electric charges create magnetic fields, while changes in magnetic fields create electricity. We use electricity in various scales, from powering computers and communication systems to commercial production systems involving electrical power equipment. Without electromagnetic radiation we would all be in the dark, for light is one of its many forms. Humans have learned to use electricity in the last couple hundred years, and in the process, electricity has transformed our civilization. Today's world uses electricity in so many ways that we cannot think to live comfortably without electricity. Electricity in nature is also important. Natural electricity causes destructive lightning flashes to our lives and properties. It is thus necessary for electrical engineering students to have solid foundation of the subject.

It gives me a great pleasure to introduce the book "Engineering Concepts of Electricity and Magnetism", written by Utpal Basu of Tata Steel. This book is an excellent presentation of the subject and, I am sure, will be found useful by the electrical engineering students of diploma and degree level in developing their concepts on basic electric and magnetic theory, and also as competent reference aid in future.

Jamshedpur
23rd. January, 2016

Dr. S K Singh
Chief of Capability Development
Tata Steel Limited

Preface

The subject Electricity and Magnetism is read with two broad perspectives by physicists and engineers. This book is intended primarily for students pursuing diploma and degree courses in electrical engineering, concentrating mainly on electrical power equipment, systems and instruments, as distinct from control and communication systems. Prerequisite to follow this book is Higher Secondary mathematics of Indian school curriculum, especially calculus, without which, as is known, electrical theory cannot be fully appreciated. The topics covered in this book, although not in strict conformance to any syllabus or semester of any examining authority, comprise part of such syllabus of any polytechnic or engineering college, under different modules. This book may be treated as a conceptual reference book and should be read in conjunction with standard text books on the subject. In fact, it was my intention to write a booklet covering brief descriptions of the various electric and magnetic quantities and their units. As the work progressed, however, I felt it would be more useful if the basic theory is covered – a subject I struggled more to conceive in my early student days than the machines, instruments and power systems later on. I have aimed to explain the basic laws and concepts of electricity and magnetism in my own way, and arranged in such a sequential order as, in my view, should be taught to students. Independent thought process beyond conventional way of learning is essential for deep insight of any subject, and this book has been written with this philosophy. 'Concept' and 'Analogy' are the key words for the book. All the concepts developed in this book will be found applicable in Electrostatics (Ch.1), and only selected ones have been copied in subsequent chapters in the form of analogy. It is, therefore, imperative, at least for beginners, to go through this first chapter minutely. Some new concepts, topics, figures and terminology will be found in various places in the book, most significant one being the marked distinction between the potential energy (PE) and stored energy (SE). Such concepts basically emerged from my own thought process, and hence, remain open for debate and corrective criticism, expected mainly from the teaching fraternity.

Inspite of my sincere care to eliminate errors, few of them might have remained, and I shall be grateful to accept and acknowledge the same.

I wish to express my sincere appreciation to my organization, colleagues and family for their continued support and encouragement, especially to my colleague Mr. BK Dubey who taught me to draw sketches in computer. I shall be failing in my duty if I do not express my gratitude to my beloved teachers who taught me the subject with care and devotion, as well as to the authors of various valuable books and papers I read both in hard and soft – too many to name. I am grateful to Dr. SK Singh, who has had authored several text books published by leading publishers, for writing foreword for this book. In fine, I am thankful to my publisher Notion Press, especially to their type setting team, for publishing the book in a good shape.

I shall deem my effort amply rewarded if the book is found useful by the students.

Jamshedpur
26th. January, 2016

Utpal Basu
ubasu58@gmail.com

Ch. 1: Electrostatics

1.1 Introduction: Electrostatics deals with static electricity, or electricity at rest. Electrostatics in Electrical Engineering is analogous to Statics in Mechanics, and Current Electricity to Dynamics. In Statics, bodies do not move under the influence of a system of forces, whereas in Dynamics bodies move. Similarly, in Electrostatics electric charges do not move continuously, whereas in Current Electricity electric charges move continuously under the influence of some potential difference.
The earliest known history of the discovery of electricity dates back to about 600 B.C., when the ancient Greek philosopher Thales of Miletus noticed that if amber (a yellow-brown resin) is rubbed with some dry substance like a piece of silk cloth, it gains the property of attracting light materials such as feather, straw, small pieces of paper, etc. For many centuries thereafter, no light was thrown on the subject. Greek name of amber is "electron" and, therefore, the source or agency that causes amber to acquire such peculiar property was first time called 'electricity' by Dr. William Gilbert, the then court physician to Queen Elizabeth, in 1600 A.D. He also showed that electricity could be produced not only in amber but also in many other substances. Such substances, when exhibiting electricity, are called 'electrified' or 'charged with electricity'. Next step of development happened in about 1733, when Charles Francois du Fay discovered that electricity appears in two kinds, *"vitreous"* (positive) and *"resinous"* (negative). In 1747, Benjamin Franklin, an American, established that electrified substances repel or attract one another depending on their being charged with similar (like) or dissimilar (unlike) electricity respectively. In 1766, Joseph Priestly (1733-1854), who is better known for the discovery of oxygen, based on the observations of Benjamin Franklin, investigated the subject further. Priestly observed that the force between two electrified substances obeys an inverse square law. This law is known today as Coulomb's Law, because it was Charles Augustin de Coulomb (1738-1806) who, by virtue of his brilliant invention of a torsion balance, was able to confirm the validity of the inverse square law in 1785 with significant accuracy.

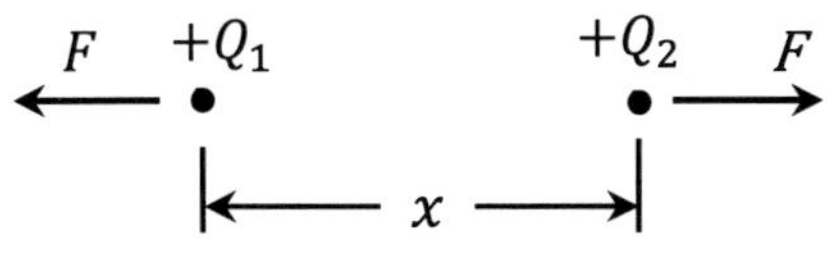

Fig.1.1(a) Coulomb's Law

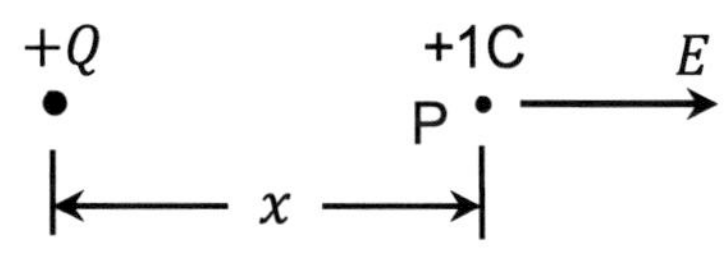

Fig.1.1(b) Field near a point charge

Fig. 1.2 Two alternative concepts behind the force between two charges

1.2 Electric Charge and Permittivity: Electric charge or static electricity is denoted by the symbol Q and is expressed in the unit coulomb (C). The size of the unit coulomb is derived from Coulomb's law that gives the force between two small (point-like) electric charges.

Referring to Fig.1.1(a), Coulombs law in Electrostatics is stated by the expression

$$F = \frac{Q_1 Q_2}{4\pi\varepsilon x^2} \quad \text{...} \quad (1.1)$$

where, F is the force in newton between the two point charges Q_1 and Q_2 coulomb and x is the distance between them in metre. The quantity ε defines a property called *permittivity* of the medium surrounding the charges. The earlier terms '*dielectric constant'* and '*specific inductive capacity*' are now obsolete. The (absolute) permittivity ε of a medium is given by the product $\varepsilon = \varepsilon_0 . \varepsilon_r$ where ε_0 is the permittivity of free space (vacuum) and ε_r is the *relative permittivity* of the medium (relative to the free-space value ε_0) surrounding the charges. Permittivity is analogous to density of matter while relative permittivity to specific gravity or relative density. The free-space permittivity ε_0 is referred to as '*primary electric constant'* having an accurate value of 8.854x10^{-12}, which is approximately $1/(36\pi \times 10^9)$ = 8.842x10^{-12} farad/metre (F/m). The numerical value of ε_0 bears a history. When the laws of electricity and magnetism were initially developed (gradually during 1785-1840), the CGS (Centimeter-Gram-Second) system of units was in existence. In this system, both the primary electric constant ε_0 and the primary magnetic constant μ_0 (Sec.3.2) were arbitrarily assigned a value of unity. Since any electric or magnetic quantity can be derived independently both from electric and magnetic considerations in terms of ε_0 and μ_0 respectively, two systems of CGS units, namely, CGS Electrostatic Units (esu) and CGS Electromagnetic Units (emu), existed. Subsequently, Maxwell's Electromagnetic Theory of Light (discovered in 1865) revealed that the permittivity ε and the permeability μ of a medium is related by the equation $\mu\varepsilon = 1/c^2$, where c is the velocity of light in that medium. This equation related esu and emu of any electric or magnetic quantity. However, the sizes of both the esu and the emu of most of the electric and magnetic quantities were found either too large or too small for practical purposes. Hence, a third system of units, Practical Units, was also in use. These developments made the unit system so complicated that the necessity of a single absolute (independent) system of units was realised. In the journey, several physicists proposed from time to time a number of systems which were either rejected for some reason or modified subsequently, finally ending with the RMKS (Rationalised Metre-Kilogram-Second) system of units or SI (International System of) units. In SI units, to retain the unit sizes of most significant electric and magnetic quantities same as those of their Practical units, derived both from electric and magnetic considerations, the value of μ_0 had to be modified from unity to 4π.10^{-7}. The value of ε_0, therefore, had also to be modified as $\varepsilon_0 = 1/(\mu_0 . c^2)$ = 8.854x10^{-12}, the free-space velocity of light c being 2.998x10^{8} m/s. With c approximated as 3x10^{8} m/s, the approximate value of ε_0 becomes

$1/(36\pi \times 10^9) = 8.842\text{x}10^{-12}$. The subject of 'rationalisation', however, is not directly related with the unit systems. In CGS systems, as well as in the (unrationalised) MKS system, the term 4π was found to occur in formulae related to systems with linear symmetry but not to systems with circular or spherical symmetry. Rationalisation merely eliminates this illogical situation. These historical developments do not interest present day students, so kept outside our present scope of study. It is sufficient now to accept and remember that $\varepsilon_0 = 8.854\text{x}10^{-12}$.

It will be seen later (Sec.1.10) that dimensionally capacitance is given by (permittivity x length), and the unit of capacitance being farad (F), permittivity is expressed in the unit of farad/metre (F/m). The relative permittivity, on the other hand, being a relative quantity is a unit-less number. Its value for free space is 1 by definition, and air is commonly taken to have the same value in place of its precise value of 1.00054. In all discussions on the subject, the medium remains implied as being the air unless mentioned otherwise. For reason that will be discussed in Sec.1.12, all substances, solids liquids and gases, have ε_r always greater than one. The insulating media in which electrostatic forces exist are also referred to as '*dielectrics*'. The exact term – insulator or dielectric – depends on the desired properties of the material for the specific application. The relative permittivity of most gaseous dielectrics, like that of air, is of the order of unity, and it varies between 2 to 10 for most solid and liquid dielectrics, except for water (pure) having a value as high as 80.

The size of unit electric charge, the coulomb, is defined from Eqn.1.1. *If two equal point charges, spaced one metre apart in vacuum, exert a force of $9\text{x}10^9$ newton on each other, the quantity of each of the charges is defined to be one coulomb.*

1.3 Electric Field Strength: The force between two charges discussed in Sec.1.2 may be thought of to occur in two alternative concepts. At any point in the space surrounding an electric charge (or a system of charges) an electric effect is experienced, and the space is said to be permeated with electric field. In other words, an electric field exists surrounding an electric charge. When two charges lie in proximity, the resulting field surrounding them is the combined effect of the individual fields of both these charges. First concept is that the force occurs by interaction between the two electric fields. Second concept is that one charge does not have any electric field to modify that of the other but experiences the force simply by virtue of its presence in the field of the other charge. In terms of lines of force (Sec.1.4), Fig.1.2 explains these two alternative concepts for similar charges. Whereas the force between electric and magnetic objects are better explained in terms of the elastic properties of their lines of forces, the second concept is conveniently applied in all definitions and derivations in terms of a '*test charge*' defined as a unit positive point charge *without* any electric field. As will be seen subsequently, the same concept also applies in Magnetism in terms of '*test pole*' (Sec.3.3), and in Electromagnetism in terms of '*test current*' (Sec.4.12).

The electric field strength, electric field intensity, or simply the electric field, E at a point in an electric field is defined as the force on a test charge placed at that point. Being merely a force, electric field strength is vector quantity. As electric field strength is force on unit charge, the force on a charge Q lying in an electric field at a point of field strength E, becomes

$$F = EQ \qquad \text{......} (1.2)$$

Hence, F may be positive or negative depending on the signs of E and Q both. The signs of E and F simply give their directions. Usual unit of electric field strength E is volt/metre (V/m), which is the same as newton/coulomb (N/C) as is given by Eqn.1.2. It will be seen later (Sec.1.8) that dimensionally electric potential (or potential difference) is given by (electric field strength x length), and the unit of potential being volt, electric field strength is expressed in volt/metre.

Referring to Fig.1.1(b), the electric field strength at a point P distant x from and due to a point charge Q is given by

$$E = \frac{Q}{4\pi\varepsilon x^2} \qquad \text{......} (1.3)$$

Some important points can be derived from Eqn.1.3. If E_0 be the field strength at some point distant x from any given amount of charge Q in vacuum, and E be the same in some dielectric medium of relative permittivity ε_r , we have

$$E_0 = \frac{Q}{4\pi\varepsilon_0 x^2} = \frac{\varepsilon_r Q}{4\pi\varepsilon_0\varepsilon_r x^2} = \frac{\varepsilon_r Q}{4\pi\varepsilon x^2} = \varepsilon_r \frac{Q}{4\pi\varepsilon x^2} = \varepsilon_r E$$

$$\text{or, } E = E_0/\varepsilon_r \qquad \text{......} (1.4)$$

Expressed in language, these relations demonstrate the following points. As $\varepsilon_r > 1$,

1) For a given amount of charge that creates the field, the field strength in any dielectric is reduced by ε_r times than that in the vacuum, or, stated otherwise,
2) To maintain the same field strength in the dielectric as in the vacuum, the quantity of the charge that creates the field has to be increased by ε_r times.

These facts will be better appreciated in Sec.1.11 on Polarization.

3) Since $E_0 > E$, let $E_0 - E = E_P$. Hence, $E_P = E_0 - E = \varepsilon_r E - E = (\varepsilon_r - 1)E$. This retarding field E_P is due to polarization of the material (Sec.1.11) and the quantity $(\varepsilon_r - 1)$ is the susceptibility of the material (Sec.1.12).

Being vector quantities electric field strengths are additive as vectors. The resultant field strength E at some point P, Fig.1.3, due to two or more charges Q_1, Q_2, etc. is given by the vector sum of the field strengths E_1, E_2, etc. at that point due to the individual charges. This is illustrated by a numerical example (Ex.1.1, Sec.1.8.1).

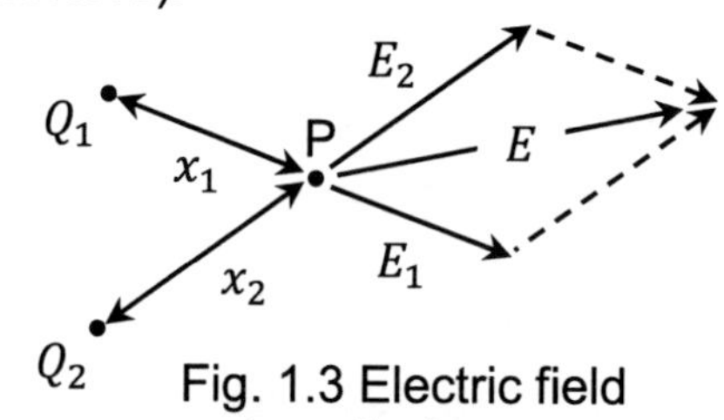

Fig. 1.3 Electric field strength due to a number of charges

1.4 Electric Flux: The line traced out by a test charge, if allowed to move freely in an electric field, is called 'a line of electric force'. Such lines of force thus originate from positive charge and terminate on negative charge, and the direction of a line of force through any point in an electric field gives the direction of the field at that point. In any electric field, there is innumerable number of such lines of force, and the profile or distribution pattern of these lines of force is referred to as the configuration of the field.

As shown in Fig.1.2(b), lines of force around a point charge are distributed radially in all directions in the entire three-dimensional space surrounding the charge, which we call the radial (configuration of the) field. Such lines of force are purely imaginary and do not have any physical existence. Concept of lines of force was first conceived and developed by the British physicist Michael Faraday to visualise the configuration of the field in a simple and convenient way. These lines of force possess certain properties in terms of which the characteristics of the field can be explained with considerable ease. Whereas electric lines of force is the qualitative aspect of electric field, '*electric flux*' is its quantitative aspect. If we arbitrarily assign some certain amount of lines of flux to be associated with a given amount of electric charge, electric field calculations become simplified to very large extent. Some exemplary cases of such simplification have been considered in Sec.1.7. In fact, we cannot think Electrical Engineering without Faraday's genius concept of electric and magnetic flux.

In RMKS system of units or SI units, unit electric flux is arbitrarily assigned to be associated with unit electric charge, that is to say, unit electric flux emanates from unit positive charge or terminates on unit negative charge. As such, electric flux is expressed in the same unit of electric charge, the coulomb, and it is denoted by Ψ. Thus, if Ψ amount of electric flux originates from or terminates on Q coulomb of electric charge, we have by arbitration

$$\Psi = Q \qquad \text{.. (1.5)}$$

However, it is not correct to say that Ψ coulomb of flux equals Q coulomb of charge, because they are two different quantities, so cannot be equated as such. Eqn.1.5 only implies that their numerical values are equal. Like electric charge Q, electric flux Ψ is a scalar quantity. It follows that the amount of flux is independent of the medium, that is, remains the same in all media. It also follows that *a closed surface of any geometry enclosing a total amount of charge Q is traversed by a total flux $\Psi = Q$.* This is the statement of **Gauss's Theorem**.

1.5 Electric Flux Density: *The electric flux density D at a point in an electric field is defined as the amount of electric flux crossing per unit area surrounding the point, the area being perpendicular or normal to the direction of the flux.* Thus, if Ψ coulomb of electric flux crosses a normal area A uniformly, the electric flux density at any point on the area is given by

$$D = \frac{\Psi}{A} \qquad \text{.. (1.6)}$$

This is the general expression of electric flux density. It is expressed in coulomb per square metre (C/m^2). It is a vector quantity, the direction being along the positive direction of the flux Ψ normal to the area A. Referring to Fig.1.4, let us consider an imaginary sphere of radius x surrounding a point charge $+Q$ at its centre. The total flux $\Psi = Q$ radiates normally and uniformly over the total surface area $A = 4\pi x^2$. Hence the flux density at any point P (on the surface of the sphere) distant x from and due to charge Q is given by

$$D = \frac{\Psi}{A} = \frac{Q}{4\pi x^2}$$

Again from Eqn.1.3, the field strength at point P

$$E = \frac{Q}{4\pi \varepsilon x^2}$$

It follows that both D and E at any point in any field have the same direction. Comparing the above two equations we get,

$$D = \varepsilon E \qquad \text{.............. (1.7)}$$

This is a general field equation that holds true for any point in any configuration of the field.

This equation shows that D and E differ by ε times, which is justified because field strength is dependent on the medium but the flux and the flux density are not. Furthermore, this equation implies that E is the cause and D is its effect.

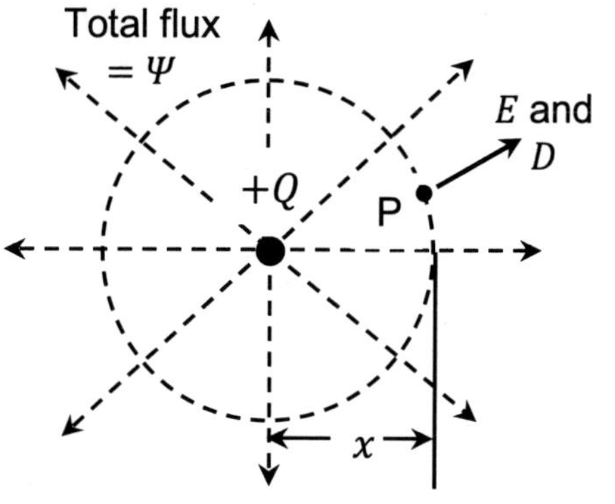

Fig.1.4 Flux and flux density around a single point charge

1.6 Electric Charge Density: We have so far been dealing with point charges or charges on small bodies only. If the size of the body is large and the material of the body is conducting, any charge imparted to the body distributes on its surface. *The electric charge density σ at any point on the surface of a charged conductor is defined as the amount of electric charge per unit surface area surrounding the point.* Hence, if Q coulomb of charge resides uniformly over a surface area A, the charge density at any point on the surface

$$\sigma = \frac{Q}{A} \qquad \text{... (1.8)}$$

Like flux density, charge density is also expressed in coulomb per square metre (C/m^2), but unlike the flux density, it is scalar quantity. The charge density at a point on the surface of a charged conductor increases with the decrease of the radius of curvature at that point. Hence, the charge density on sharp points or on edges is greater than that over flat area. This is the reason the top of lightning conductor on buildings is made pointed. The sphere being the only geometry that has a surface of uniform radius of curvature R, the charge density at all points on the surface of a conducting sphere has a uniform value of $Q/(4\pi R^2)$.

Consider now a point on the surface of a charged conductor of any geometry. If σ is charge density at that point, flux density at that point $D = \sigma$, and electric field strength at that point $E = D/\varepsilon = \sigma/\varepsilon$, where ε is the permittivity of the medium outside the conductor. This constitutes **Coulomb's Theorem**, which states that *Electric field strength at a point on the surface of a charged conductor is σ/ε, where σ is the charge density at that point and ε is the permittivity of the medium outside the conductor.*

The relation $D = \sigma$ states that the magnitude of the flux density vector D equals the numerical value of the charge density scalar σ.

1.7 Some Common Field Derivations:

1.7.1 Field near charged straight conductor: In Fig.1.5, P is a point distant x from a straight thin long conductor carrying uniformly distributed charge q per unit length, and it is required to derive the electric field strength at the point P due to the charged conductor.

(a) By Coulomb's law:

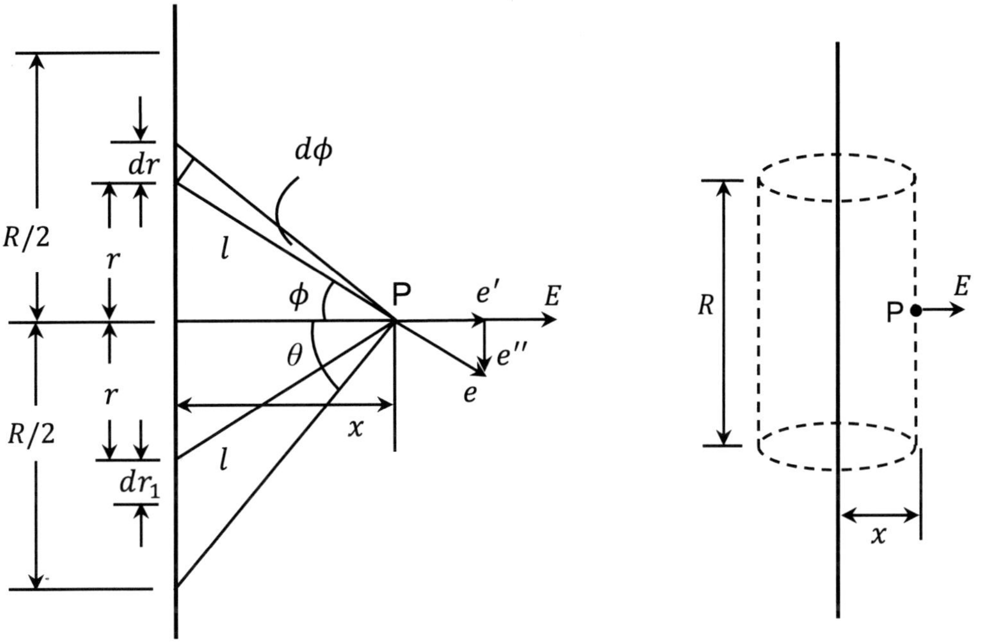

(a) By Coulomb's Law (b) By flux concept

Fig.1.5 Field near charged straight conductor

Referring to Fig.1.5(a), let us first consider a finite length R of the conductor, the point P being distant x from the mid-length of the conductor, as shown. Consider now an elemental length dr of the conductor, distant l from the point P. Field at P due to the charge $q.dr$ on the length dr, $e = \frac{q.dr}{4\pi\varepsilon l^2}$. It can be resolved into its horizontal component $e' = e.Cos\,\phi$ and vertical component $e'' = e.Sin\,\phi$. Considering a similar elemental length dr_1 of the conductor on the other side of dr, as shown, the horizontal component of the field due to dr_1 at point P is the same, both in magnitude and direction, as that due to dr. Hence, these two horizontal components are additive algebraically. The corresponding vertical components of the field at point P due to dr and dr_1 are also same in magnitude but opposite in direction, so cancel out. The total electric field at P due to the entire length R of the conductor is thus given by

$$E = 2\int_{r=0}^{r=R/2} e' = 2\int_{r=0}^{r=R/2} e.Cos\,\phi = 2\int_{r=0}^{r=R/2} \frac{q.dr.Cos\,\phi}{4\pi\varepsilon l^2}$$

Now, $dr.Cos\,\phi = l.d\phi$ and $Cos\,\phi = x/l$, or, $\frac{dr}{l} = \frac{d\phi}{Cos\,\phi} = \frac{d\phi}{x/l}$ or, $\frac{dr}{l^2} = \frac{d\phi}{x}$

Therefore, $E = 2\int_{\phi=0}^{\phi=\theta} \frac{q.Cos\,\phi.d\phi}{4\pi\varepsilon x}$, where $Sin\,\theta = \frac{R/2}{\sqrt{(R/2)^2+x^2}}$

Or, $E = \frac{2q}{4\pi\varepsilon x}\int_0^\theta Cos\,\phi\,.\,d\phi = \frac{q}{2\pi\varepsilon x}\,[Sin\phi]_0^\theta$

$$= \frac{q}{2\pi\varepsilon x} Sin\theta = \frac{q}{2\pi\varepsilon x}\left[\frac{R/2}{\sqrt{(R/2)^2+x^2}}\right] \quad \text{.................. (1.9(a))}$$

When the length of the conductor R is very great compared to the distance x, $(R \gg x)$, $\theta = \pi/2$, so $Sin\,\theta$ (or the term within the square brackets) = 1, and then

$$E = \frac{q}{2\pi\varepsilon x} \quad \text{................................ (1.9)}$$

Eqn.1.9 may also be obtained directly if the integration is performed with the limits $r = 0$ to $r = R/2$, or $\phi = 0$ to $\pi/2$, which is equivalent to the condition $R \gg x$. The flux density at P, when $R \gg x$,

$$D = \varepsilon E = \frac{q}{2\pi x} \quad \text{...................................... (1.10)}$$

(b) By flux concept: Referring to Fig.1.5(b), consider an imaginary cylinder of length R and radius x with its axis along the charged conductor, such that the point P lies on the curved surface of the cylinder. The conductor being very long compared to the distance x, the flux distribution surrounding the conductor is radial and uniform over the curved surface of the cylinder. The charge on length R of the conductor, $Q = qR$, and the flux originating from this charge, $\Psi = Q = qR$ crosses the curved surface area of the cylinder $A = 2\pi xR$ normally and uniformly. As such, the flux density at any point like P on the curved surface of the cylinder is $\Psi/A = qR/(2\pi xR) = q/(2\pi x)$, which is Eqn.1.10. Hence the field at P, $E = D/\varepsilon = q/(2\pi\varepsilon x)$, which is Eqn.1.9.

1.7.2 Field between two oppositely charged parallel conducting plates: In Fig.1.6, A and B are two thin parallel conducting plates of area A each and separated by a small distance d. Neglecting the earth capacitances of the plates, as will be discussed in Sec.1.10.5, the charges $\pm Q$ imparted to the plates are mutually attracted to reside on their facing or inner active sides and get distributed over the area A, creating an electric field in the dielectric between the plates. Neglecting further the "fringing effect" near the edges of the plates as described below, the charge density over the plates is of uniform value $\sigma = Q/A$. We now derive the field strength at some point P in the space between the plates.

(a) By Coulomb's law: Let us first consider that the plates are circular, each of radius R, and coaxial, and the point P lies on their common axis at a distance x from plate A, as shown in the figure. The field at P due to the charge on the elemental area $ds.dr$ on the elemental ring shown on plate A, $e = \frac{\sigma.ds.dr}{4\pi\varepsilon l^2}$. Over the entire elemental ring of radius r and width dr, the vertical components of all such elemental fields mutually cancel out, and the total horizontal component becomes

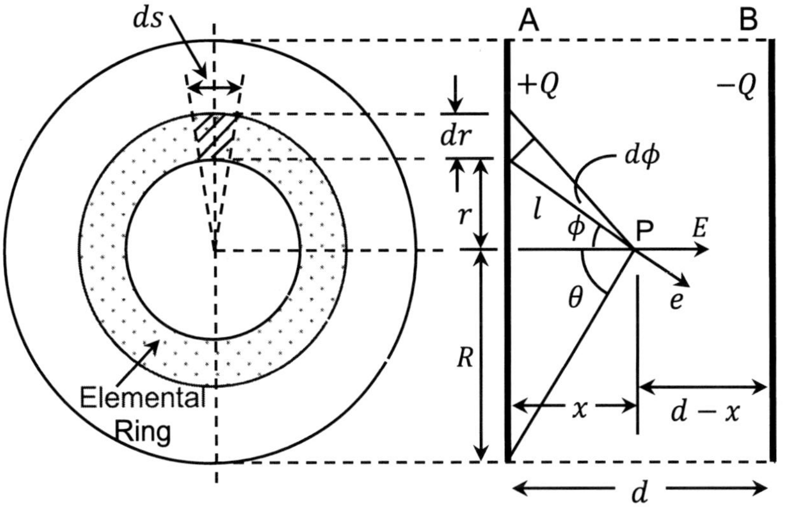

Fig.1.6 Field between two charged parallel conducting plates by Coulomb's Law

$$dE' = \int_{s=0}^{s=2\pi r} e.\,Cos\,\phi = \frac{\sigma.dr.Cos\,\phi}{4\pi\varepsilon l^2}\int_0^{2\pi r} ds = \frac{\sigma}{2\varepsilon}.\frac{r.dr.Cos\,\phi}{l^2}$$

Field at point P due to the entire charge $+Q$ on the positive plate A

$E' = \int_{r=0}^{r=R} dE' = \frac{\sigma}{2\varepsilon}\int_0^R \frac{r.dr.Cos\,\phi}{l^2}$ Now, $dr.\,Cos\,\phi = l.\,d\phi$ and $Sin\,\phi = r/l$.

So, $\frac{dr.Cos\,\phi}{l} = d\phi$ or, $\frac{r.dr.Cos\,\phi}{l^2} = \frac{r}{l}d\phi = Sin\,\emptyset\,.\,d\phi$

Therefore, $E' = \frac{\sigma}{2\varepsilon}\int_{\emptyset=0}^{\emptyset=\theta} Sin\,\phi\,.\,d\phi$, where $Cos\,\theta = \frac{x}{\sqrt{x^2+R^2}}$

$$= \frac{\sigma}{2\varepsilon}\,[-Cos\phi]_0^\theta = \frac{\sigma}{2\varepsilon}[1 - Cos\,\theta]$$

$$= \frac{\sigma}{2\varepsilon}\left[1 - \frac{x}{\sqrt{x^2+R^2}}\right]$$

This field due to the charge $+Q$ on plate A is repulsive one. Plate B with charge $-Q$ creates an attractive field $E'' = \frac{\sigma}{2\varepsilon}\left[1 - \frac{d-x}{\sqrt{(d-x)^2+R^2}}\right]$ at the point P. Total field at P due to the charges on both the plates, therefore,

$$E = E' + E'' = \frac{\sigma}{2\varepsilon}\left[2 - \frac{x}{\sqrt{x^2+R^2}} - \frac{d-x}{\sqrt{(d-x)^2+R^2}}\right]$$

[**Fringing effect**: On the surface of any plate $x = 0$, so

$$E = \frac{\sigma}{2\varepsilon}\left[2 - \frac{d}{\sqrt{d^2+R^2}}\right]$$

and at the mid distance between the plates $x = d/2$, so

$$E = \frac{\sigma}{2\varepsilon}\left[2 - \frac{d}{\sqrt{(d/2)^2+R^2}}\right]$$

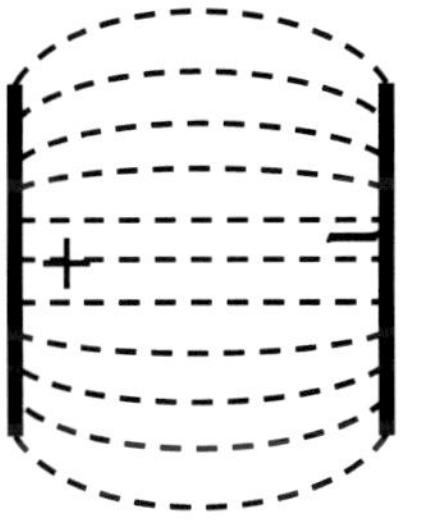

Fig.1.7 Fringing effect

These expressions show that field strength on the surfaces of the plates is the maximum and decreases gradually to some minimum at the mid distance between the plates. This results in diverging or bulging out of the lines of force from both the plates towards their mid distance, as shown at Fig.1.7. This effect is referred to as *'fringing'*, which decreases with decrease of the ratio d/A. Again, for a given value this ratio, the fringing effect increases from the center of the plates towards its edges. When this ratio is small, the fringing effect practically remains confined to the edges of the plates only. As such, the fringing effect is also referred to as *'edge effect'*. Thus, the fringing effect increases the effective sectional area of the air gap, and so reduces the air gap flux density for a given amount of flux. The effect also occurs in Magnetism (Sec.3.7.1) and Electromagnetism (Sec.4.7.4), where it becomes too significant in many practical situations to ignore, and so has to be accounted for.]

When the area of the plates is very great compared with their distant apart, $R \gg d$, the term within the square brackets approaches 2, and then

$$E = \frac{\sigma}{\varepsilon} = \frac{Q}{\varepsilon A} \quad \text{.. (1.11)}$$

Eqn.1.11 could also be obtained directly if the integration were performed with the limits $r = 0$ to ∞, or $\phi = 0$ to $\pi/2$, which is equivalent to the condition $R \gg d$. The condition $R \gg d$ simply implies that the "fringing effect" is negligible, so the field between the plates is parallel or uniform.

It is evident that when Eqn.1.11, that does not contain the term x, applies,

(a) the geometry of the plates does not bear any significance, that is, need not necessarily be circular but may be of any other shape, and as such the condition $R \gg d$ assumes the general form $A \gg d$, and also that

(b) the location of the point P may be anywhere between the plates, not necessarily on the axis. This is also logically justified, as the field is uniform.

The summary of the above derivations is that on neglecting the fringing effect ($A \gg d$), *the electric field strength at any point in the space between two close parallel conducting plates with charges $\pm Q$, due to the charge on any one plate is $Q/(2\varepsilon A)$, and so the field due to both the plates is $Q/(\varepsilon A)$*, which is Eqn.1.11.

Electric flux density at any point between and due to the two plates, when $A \gg d$,

$$D = \varepsilon E = \sigma = Q/A \quad \text{...................................... (1.12)}$$

(b) By flux concept: Total electric flux from the positive plate to the negative plate $\Psi = Q$. As the field between the plates is uniform, the flux density at any point like P between the plates $D = \Psi/A = Q/A = \sigma$, which is Eqn.1.12. The field at P, $E = D/\varepsilon = Q/(\varepsilon A) = \sigma/\varepsilon$, which is Eqn.1.11.

1.7.3 Field outside charged spherical conductor: In Fig.1.8, the charge $+Q$ imparted to a spherical conductor of radius R resides on the outer surface of the sphere with uniform density $\sigma = Q/(4\pi R^2)$, and creates a uniformly distributed radial field in the space surrounding the sphere. We are to obtain an expression for the field strength at some point P outside the sphere, distant x from its centre.

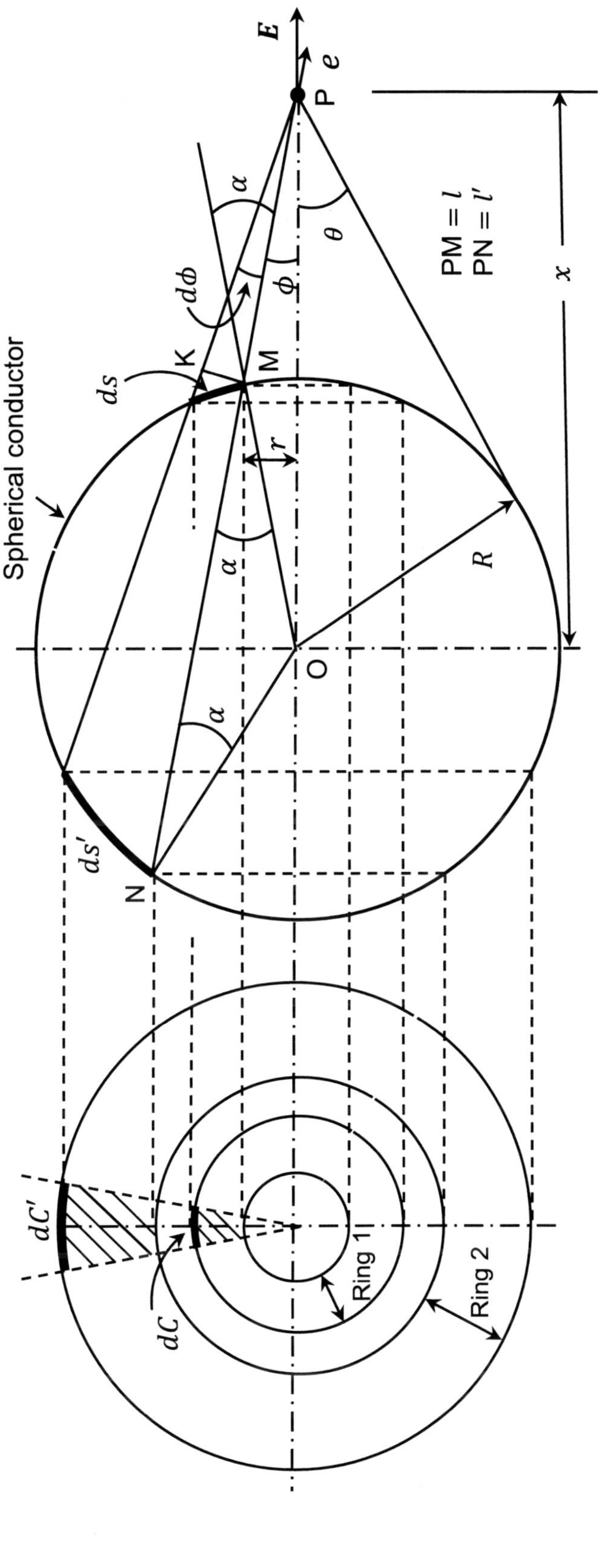

Fig.1.8 Field outside charged spherical conductor by Coulomb's Law

(a) By Coulomb's law: The electric field at P due to the charge on the elemental area $ds.dc$ on the elemental ring 1, $e = \dfrac{\sigma.ds.dc}{4\pi\varepsilon l^2}$

and that due to the charge on the elemental area $ds'.dc'$ on the elemental ring 2,

$$e' = \frac{\sigma.ds'.dc'}{4\pi\varepsilon(l')^2}$$

both acting in the same direction, and so are additive algebraically. Again, both the elemental areas subtend the same solid angle at the point P

$$d\omega = \frac{ds.dc.Cos\alpha}{l^2} = \frac{ds'.dc'.Cos\alpha}{(l')^2} \quad \text{or,} \quad \frac{ds.dc}{l^2} = \frac{ds'.dc'}{(l')^2} \quad \text{and so,} \quad e = e'$$

This condition being applicable to all such pairs of elemental areas on the two elemental rings, the field due to both the rings will be twice the field due to ring 1.
Over the entire ring 1, the vertical components of all such elemental fields like e mutually cancel out, and the total horizontal component

$$dE = \int_{c=0}^{c=2\pi r} e.Cos\,\phi = \frac{\sigma.ds.Cos\phi}{4\pi\varepsilon l^2}\int_{c=0}^{c=2\pi r} dc = \frac{\sigma.ds.Cos\phi}{4\pi\varepsilon l^2}2\pi r = \frac{\sigma}{2\varepsilon}.\frac{r.ds.Cos\phi}{l^2}$$

Now, the length MK is $ds.Cos\,\alpha = l.d\phi$, and $Sin\,\phi = r/l$. So, $\dfrac{ds}{l} = \dfrac{d\phi}{Cos\,\alpha}$

$$\text{or,} \quad \frac{ds.r}{l^2} = \frac{r}{l}\frac{d\phi}{Cos\,\alpha} = \frac{Sin\phi.d\phi}{Cos\,\alpha} \quad \text{and so,} \quad dE = \frac{\sigma}{2\varepsilon}\frac{Sin\phi\ Cos\phi}{Cos\,\alpha}d\phi$$

To obtain α in terms of ϕ, we consider the triangle OPM, wherein

$$\frac{R}{Sin\,\phi} = \frac{x}{Sin(\pi-\alpha)} = \frac{x}{Sin\,\alpha} \quad \text{or,} \quad Sin\,\alpha = \frac{x}{R}Sin\,\phi$$

or, $Cos\,\alpha = \sqrt{1 - Sin^2\alpha} = \sqrt{1-(x/R)^2 Sin^2\phi}$. And, $dE = \dfrac{\sigma}{2\varepsilon}\dfrac{Sin\phi.Cos\,\phi}{\sqrt{1-(x/R)^2\,Sin^2\phi}}d\phi$

Field at P due to the entire charge Q on the sphere

$$E = 2\int_{\phi=0}^{\phi=\theta} dE \ \text{(where } Sin\,\theta = R/x\text{)} = \frac{\sigma}{\varepsilon}\int_{\phi=0}^{\phi=\theta}\frac{Sin\phi.Cos\,\phi}{\sqrt{1-(x/R)^2\,Sin^2\phi}}d\phi$$

$$= \frac{\sigma}{\varepsilon}\left[-(R/x)^2\sqrt{1-(x/R)^2 Sin^2\phi}\right]_{\phi=0}^{\phi=\theta} = \frac{\sigma}{\varepsilon}.\frac{R^2}{x^2} = \frac{Q}{4\pi R^2}.\frac{R^2}{\varepsilon x^2}$$

$$\text{or, } E = \frac{Q}{4\pi\varepsilon x^2} \qquad \text{.......... (1.13)}$$

as if the distributed charge Q on the sphere is concentrated at its centre.
For a point on the surface of the sphere, $x = R$, and

$$E = \frac{Q}{4\pi\varepsilon R^2} \qquad \text{.......... (1.14)}$$

Electric flux density at point P

$$D = \varepsilon E = \frac{Q}{4\pi x^2} \qquad \text{.......... (1.15)}$$

And that on the surface of the sphere

$$D = \varepsilon E = \frac{Q}{4\pi R^2} \qquad \text{.......... (1.16)}$$

Eqns.1.13 to 1.16 derive the important conclusion that *for any point outside a charged spherical conductor the total charge distributed over the entire surface of the sphere behaves as if concentrated at its centre.*

(b) By flux concept: There being no charge inside the sphere, the total flux $\Psi = Q$ radiates outside the sphere. Considering an imaginary sphere of radius x concentric with the charged sphere, as shown in Fig.1.9, flux density at any point like P on the surface of this imaginary sphere $D = \Psi/(4\pi x^2) = Q/(4\pi x^2)$ which is Eqn.1.15, and the field at P, $E = D/\varepsilon = Q/(4\pi\varepsilon x^2)$ which is Eqn.1.13.

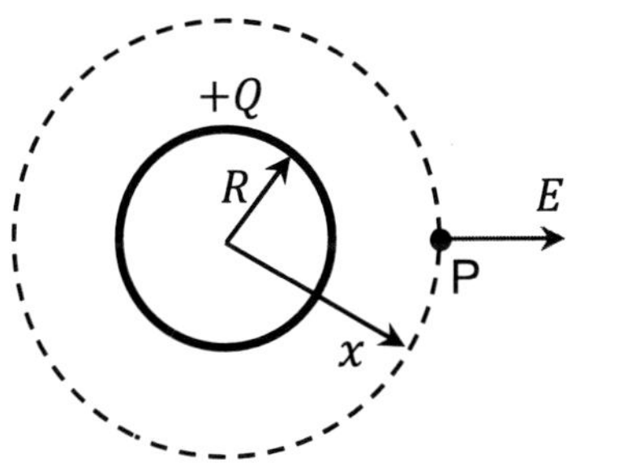

Fig.1.9 Field outside charged spherical conductor by flux concept

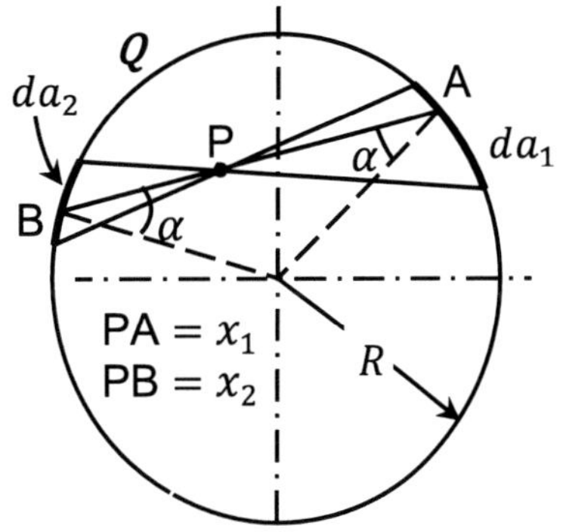

Fig.1.10 Field inside charged spherical conductor by Coulomb's Law

1.7.4 Field inside charged spherical conductor: In Fig.1.10, the charge $+Q$ imparted to the spherical conductor of radius R resides on the surface of the sphere with uniform density $\sigma = Q/4\pi R^2$, and we are to evaluate the field strength at some point P inside the sphere.

(a) By Coulomb's law: Considering two elemental areas da_1 and da_2 on the surface of the sphere and on opposite sides of P, as shown, the field at P due to the charge on area da_1, $e_1 = \dfrac{\sigma.da_1}{4\pi\varepsilon {x_1}^2}$ and that due to the charge on area da_2, $e_2 = \dfrac{\sigma.da_2}{4\pi\varepsilon {x_2}^2}$ where ε is the permittivity of the medium inside the sphere. These two fields act in opposition. Again, the two areas da_1 and da_2 subtend the same solid angle at point P

$$d\omega = \frac{da_1.Cos\,\alpha}{{x_1}^2} = \frac{da_2.Cos\,\alpha}{{x_2}^2} \quad \text{or,} \quad \frac{da_1}{{x_1}^2} = \frac{da_2}{{x_2}^2} \quad \text{and so, } e_1 = e_2$$

Hence e_1 and e_2 mutually cancel out. This cancellation applies to all such pairs of opposite elemental areas over the entire surface of the sphere. The field inside the sphere is thus zero.

(b) By flux concept: Since any charge imparted to a conductor (of any shape) resides on its outer surface, and all the flux emanates from or terminates on the outer surface, there is no flux, inside the conductor. As such, flux density and field strength at any point inside any charged conductor is zero. From the radial and parallel field configurations considered above, it can be appreciated how much simplification has been achieved in the treatment by the concept of electric flux, as was mentioned earlier in Sec.1.4.

1.7.5 Field near short electric dipole:

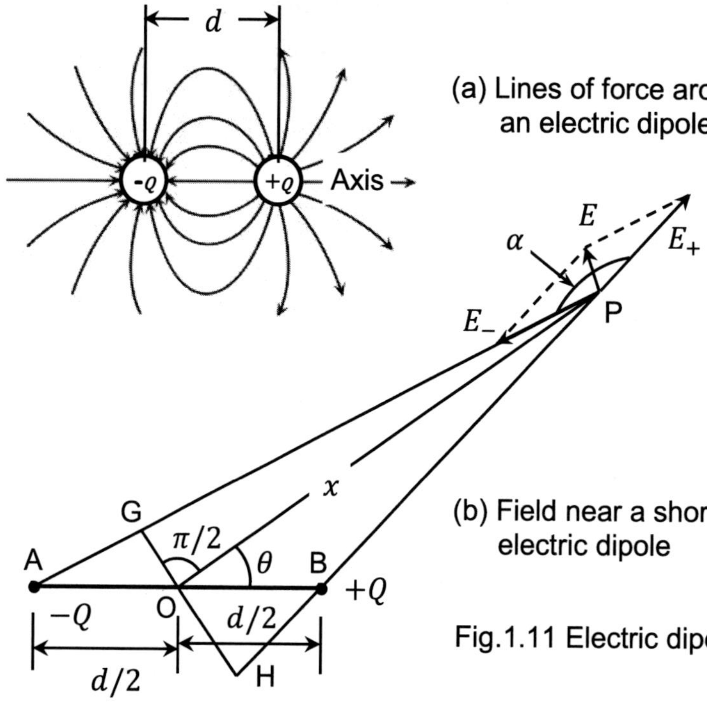

(a) Lines of force around an electric dipole

(b) Field near a short electric dipole

Fig.1.11 Electric dipole

Two equal and opposite point charges $+Q$ and $-Q$ separated by a distance d constitute an '*electric dipole*'. Electric field lines surrounding a dipole are shown at Fig.1.11(a). Axis of the dipole is the line passing through the points of the charges. Let us now derive the electric field strength E at any point P distant x from the centre of the dipole, Fig.1.11(b). The field at P due to $+Q$ is given by $E_+ = Q/[4\pi\varepsilon(\text{PB})^2]$, and that due to $-Q$, $E_- = Q/[4\pi\varepsilon(\text{PA})^2]$, their directions being as shown. The resultant field is E. The direction of E is, of course, the direction of the line of force at and through the point P. Line GH is ⊥ to line PO. The dipole being short, x is very great compared to d, that is, $x \gg d$, and as such, the angles at G and H can be approximated as $\pi/2$ each, so that PG = PH $= x$. With this approximation, both the angles GAO and HBO are θ, so that GA=HB=$(d/2)Cos\theta$. Hence,

PB=PH−HB=$(d/2)\ Cos\theta,$ and PA=PG+GA=$x + (d/2)\ Cos\theta$.

Putting these values of PB and PA, we get

$$E_+ = \frac{Q}{4\pi\varepsilon[x-(d/2)Cos\theta]^2} \quad \text{and} \quad E_- = \frac{Q}{4\pi\varepsilon[x+(d/2)Cos\theta]^2}$$

The angle between E_+ and E_- being α, the magnitude of E is given by

$$E = \sqrt{E_+^{\,2} + E_-^{\,2} + 2.E_+.E_-.Cos\,\alpha}$$

With the approximation $x \gg d$, since the angles at G and H are $\pi/2$ each, the angle α seems to be very close to π, and $Cos\,\pi = -1$. Hence, if we approximate α as π, the magnitude of E is evaluated as

$$E = \sqrt{E_+^2 + E_-^2 - 2.E_+.E_-} = \sqrt{(E_+ - E_-)^2} = E_+ - E_-$$

$$= \frac{Q}{4\pi\varepsilon}\left[\frac{1}{[x-(d/2)Cos\theta]^2} - \frac{1}{[x+(d/2)Cos\theta]^2}\right] = \frac{Q}{4\pi\varepsilon}\frac{2xdCos\theta}{[x^2-\{(d/2)\,Cos\theta\}^2]^2}$$

As $x \gg d$, the term within the square brackets in the denominator reduces to x^4, and

$$E = \frac{Q}{4\pi\varepsilon}\frac{2dCos\theta}{x^3} = \frac{Qd}{2\pi\varepsilon}\frac{Cos\theta}{x^3} \qquad \text{.............................. (1.17(a))}$$

This equation has a serious limitation. For points on the axis of the dipole, we have $\theta = 0,\ Cos\theta = 1$, and E becomes

$$E = \frac{Qd}{2\pi\varepsilon x^3} \qquad \text{.................................... (1.17)}$$

This is correct, because on the axis of the dipole $\alpha = \pi$ exactly. But for points on the perpendicular bisector plane of the dipole, $\theta = \pi/2$, $Cos(\pi/2) = 0$, so E becomes zero, which is incorrect. This is in fact the effect of approximating α as π. As θ increases from zero, the degree of inaccuracy in Eqn.1.17(a) gradually increases, and finally at $\theta = \pi/2$ the equation becomes invalid. It can be shown that for points on the perpendicular bisector plane, under the condition $x \gg d$, the field is given by $E = \frac{Qd}{4\pi\varepsilon x^3}$. It should be noticed that the field strength near an electric dipole varies as $(1/x^3)$, whereas that near a point charge as $(1/x^2)$.

1.8 Electric Potential, Potential Difference and Potential Energy: As we know, Potential Energy (PE) is any type of stored energy. It can be gravitational, elastic, pressure, chemical, nuclear, electrostatic, magnetic, etc. On the other hand, Kinetic Energy (KE) is associated with any sort of motion, like mechanical motion, vibration, electric current (motion of electrons and ions), etc. Any radiant energy like heat, light or sound involves motion of photons and is thus kinetic energy. Electric potential (or potential difference, abbreviated like PD and p.d.) is a measure of electrostatic potential energy. It is denoted by the symbol V. *Electric potential at a point in an electric field is defined as the amount of work done by some external agency in moving a test charge against the field from some arbitrarily chosen zero reference point to the specified point.* The selection of zero reference point of electric and magnetic potential (and potential energy) requires an explanatory note. As we know, potential energy is always stated with reference to some zero level. In Mechanics, the zero reference point of gravitational potential energy of a mass is chosen as the ground level or the earth surface. For electric and magnetic potential, the zero reference point is, in general, where the force of the field on the test charge, test pole, or test current is zero. This may be infinite distance if the field extends up to infinity. In some situations, the field does not extend up to infinity, and in such cases some equivalent zero reference is defined. Even if the field extends up to infinity, the zero reference may also occur at some point other than infinity. We will gradually come across such variety of situations. Unless specified otherwise, however, the zero reference point of electric and magnetic potential remains implied as infinite distance. Electric or magnetic potential or potential energy is thus always relative.

However, whatever zero reference we may like to choose, the change in the potential (PD) between two points, or the change in the potential energy of an electrified or magnetised object with change in its position, remains the same. In fact, the absolute values (with respect to any zero reference) of electric and magnetic potential and potential energy are not so significant; rather their differences or changes with their positions will be found to be of sole importance in subsequent discussions.

Electric field strength E at some point in an electric field being the force on a test charge at that point, in the general case that E varies with distance x, the work done *by* the field in moving the charge over a very short distance dx along the direction of the field is $E.dx$. So, work done *against* the field is $-E.dx$. Potential or PD being the work done *against* the field, it is given by the general form of the integral

$$V = -\int E.dx \qquad \text{......} (1.18)$$

with limits of x along the direction of $(-E)$. Potential of a point may also be negative with respect to some zero reference point. Eqn.1.18 is in integral form; its differential form is $dV = -E.dx$, or $E = -dV/dx$. This simply implies that if E acts along positive direction of x, V decreases with increase in x along positive direction. Referring to Fig.1.12, the slope of the potential line, $tan\,\alpha = dV/dx$, gives the distance rate of *rise* of the potential at a point of field strength E and potential V, and is called the '**Potential Gradient**' of the field at that point.

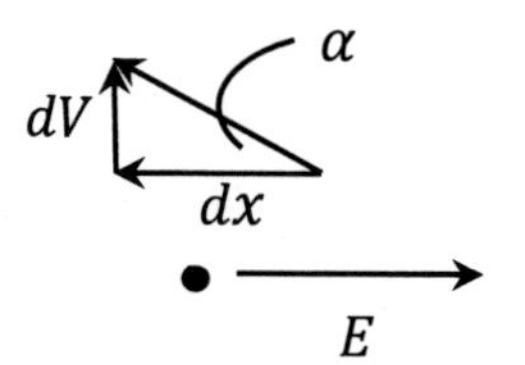

Fig.1.12
Potential Gradient

Potential difference (PD) between two points 2 and 1 in an electric field, denoted by V_{21}, *is defined as the amount of work done against the field in moving a test charge from point 1 to point 2.* Thus, V_{21} denotes the potential of point 2 with respect to, or relative to, that of point 1. The PD is, therefore, *relative potential*, as distinct from the *absolute potential* with respect to some stated or implied zero reference.

$$V_{21} = -\int_1^2 E.dx \qquad \text{......} (1.19)$$

This is the general expression of PD. It follows that $V_{21} = -\int_1^2 E.dx = \int_2^1 E.dx = -V_{12}$. Hence, like potential the PD may also be positive or negative.

The mathematics $V_{BA} = -\int_A^B E.dx = \int_B^A E.dx$ also asserts that the potential or PD may be evaluated as the work done *by* the field. This is obvious because the amount of the work done by a force along its direction is the same as the amount of work done against the force in the reverse direction. Hence, the potential or PD remains the same no matter whether computed as work done against the field or by the field. It follows that electric potential or PD, in effect, specifies some point (or position) in an electric field with respect to some reference point.

Electric potential or PD is expressed in the unit of volt (V). In fact, the unit 'volt' is so common that it is customary to mean the quantity electric potential or PD itself by the term 'voltage' derived from its unit 'volt'. As electric potential or PD represents work, volt is joule/coulomb (J/C), like the field strength described earlier as newton/coulomb (N/C). Eqn.1.18 explains that the unit of electric field strength is volt per metre.

Any quantity, or the size of any unit, may be defined in a number of ways depending on the relationship or equation used for expressing the quantity. Let us elaborate this understanding with the example of electric potential. As electric potential or PD is the amount of work done per unit charge in an electric field, the PD between two points in an electric field is defined as one volt if one joule of work is required to move one coulomb of charge from one point to another. Again, using Eqn.1.18, potential or PD may be defined as the work done in moving unit charge through unit distance in a uniform electric field, and accordingly, the volt may also be defined as the PD between two points one metre apart in a uniform field of one newton/coulomb.

The amount of work done by some external agency in moving an electric charge (more specifically, in moving the body carrying the charge) against an electric field from some reference point of potential is stored with the charge as its electrostatic potential energy. Let us designate it as PE_E. The subscript E denotes 'electric', as exactly the same concept applies to the corresponding magnetic case, Sec.3.8. The PE_E, therefore, involves two independent agencies, an electric field and an electric charge, and it must be expressed with respect to some stated or implied reference point. If a charge Q lies in an electric field at some point of potential V (*created by some other charge or system of charges*), the electrostatic potential energy of the charge Q at that point, with respect to the zero reference point of potential, becomes

$$PE_E = QV \qquad \text{.. (1.20)}$$

Hence, PE_E may be positive or negative depending on the signs of Q and V both. If V be the potential difference V_{21} in Eqn.1.20, then the product $Q.V_{21}$ gives the electrostatic potential energy $(PE_E)_{21}$ of charge Q at point 2 with respect to zero reference at point 1. Thus, if a charge Q moves against a positive potential difference V, its electrostatic potential energy PE_E increases by the amount

$$\Delta PE_E = QV$$

Electric potential or PD may, therefore, be regarded as 'specific electrostatic PE' possessed by electric charge, means, the electrostatic PE possessed by unit positive charge.

Being work done, electric potential, PD and potential energy are all scalar quantities, and so are additive algebraically.

1.8.1 Potential and potential energy in the field of point charge: Referring to Fig.1.13(a), the potential of a point P distant x from and due to a point charge $+Q$ is computed as

$$V = -\int_{x=\infty}^{x=x} E.dx = -\frac{Q}{4\pi\varepsilon}\int_{\infty}^{x}\frac{1}{x^2}.dx$$

$$= -\frac{Q}{4\pi\varepsilon}\left[-\frac{1}{x}\right]_{x=\infty}^{x=x} = -\frac{Q}{4\pi\varepsilon}\left[-\frac{1}{x}\right]$$

$$= \frac{Q}{4\pi\varepsilon x} \qquad \text{.............................. (1.21)}$$

This is positive and is given by the positive shaded area in Fig.(a), also by the height 12 in Fig.(b). These positive curves in Figs.(a) and (b) also represent, to different scales, the force F on, and the PE_E of, some positive point charge at point P respectively. The negative area and curves apply for some negative charge at point P. Like $E = -dV/dx$, we have $F = -dPE_E/dx$, which implies that when F acts along positive direction of x, PE_E decreases with increase in x along positive direction.

It is natural law that any body or system always tends to move to the lowest possible level of potential by releasing its PE. Flowing of water to lower level is, perhaps, the simplest example. Electric and magnetic bodies and systems are also no exception to this universal rule. Thus, in Fig.1.13(a), if the test charge at the point P, with $PE_E = +V$, were released to move, it would move to some infinite distance where the lowest potential is zero, by releasing its PE_E by V (from V to zero). However, the lowest potential may not always be zero, but may be negative also. This happens, for example, if the field-creating charge Q in Fig.1.13(a) were negative $(-Q)$. In this case, the test charge would move from some infinite distance of zero potential to the point P of negative potential $-V$, by releasing its PE_E by V (from zero to $-V$).

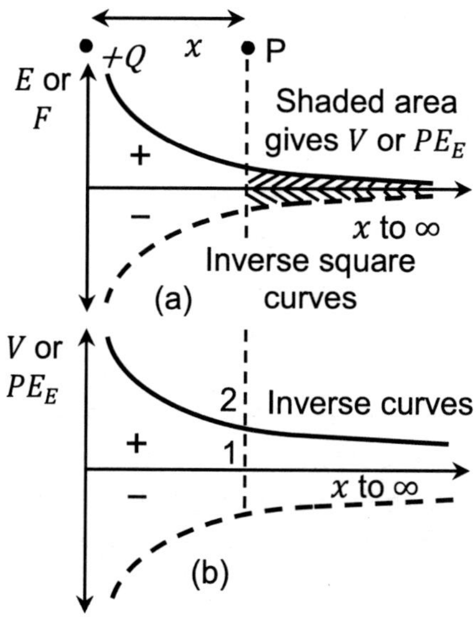

Fig.1.13 E, F, V and PE_E due to point charge

Example 1.1: *In Fig.1.14, ABCD is a square of 15 cm side. Find the electric field strength, electric flux density and electric potential at point D due to charges +10 pC, -4 pC and +10 pC at points A, B and C respectively, the medium being air.*

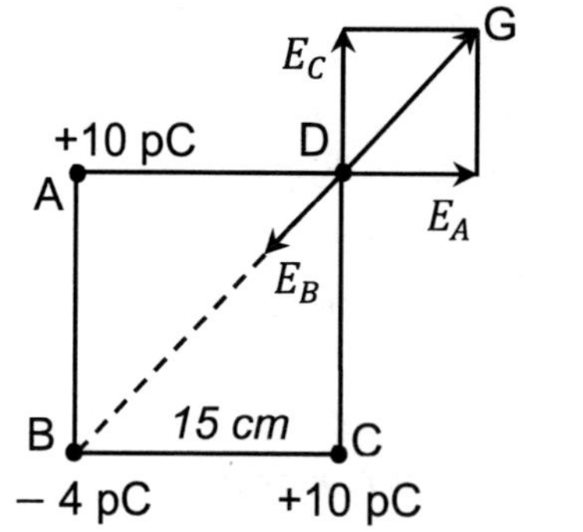

Fig.1.14 Numerical example

Solution: Field strength at D due to charge at A

$$E_A = \frac{10.10^{-12}}{4\pi(8.854.10^{-12}).(15.10^{-2})^2} = 4 \text{ V/m}$$

Similarly, field strength at D due to charge at C,

$$E_C = 4 \text{ V/m}$$

These two fields add vectorially to give $\overrightarrow{DG} = \sqrt{2}.4 =$ 5.656 V/m.

The field at D due to charge at B, $E_B = \dfrac{4.10^{-12}}{4\pi\left(8.854.10^{-12}\right).\left(\sqrt{2}.15.10^{-2}\right)^2} =$ 0.8 V/m.

As E_B is just in opposition to $\overrightarrow{DG}$, the resultant field strength at D due to all the three charges, $E_D =$ 5.656 – 0.8 = 4.856 V/m, directing along $\overrightarrow{BD}$.

Flux density at point D, $D_D = \varepsilon_0 E_D = 8.854.10^{-12}.4.856 = 43$ pC/m^2 directing along $\overrightarrow{BD}$.
The potentials at D due to the charges at A, B and C are

$$V_A(= V_C) = \frac{10.10^{-12}}{4\pi(8.854.10^{-12}).15.10^{-2}} = 0.6 \text{ V}, \; V_B = \frac{-4.10^{-12}}{4\pi(8.854.10^{-12}).\sqrt{2}.15.10^{-2}} = -0.17 \text{ V}.$$

Hence, the net potential at D, V_D = 0.6 + 0.6 – 0.17 = 1.03 V.

1.8.2 Potential difference in the field of point charge: Let us consider the PD between two points 2 and 1 in the electric field due to a positive point charge $+Q$. Two situations may arise. The two points may be collinear with the charge (collinear configuration); else, they may not be (offset configuration). We consider the two cases separately.

(a) Collinear configuration: In Fig.1.15(a), it is simply computed as

$$V_{21} = -\int_{x_1}^{x_2} E.dx = -\frac{Q}{4\pi\varepsilon}\int_{x_1}^{x_2}\frac{1}{x^2}dx = \frac{Q}{4\pi\varepsilon}\left[\frac{1}{x}\right]_{x_1}^{x_2} = \frac{Q}{4\pi\varepsilon x_2} - \frac{Q}{4\pi\varepsilon x_1} = V_2 - V_1,$$

which is obvious.

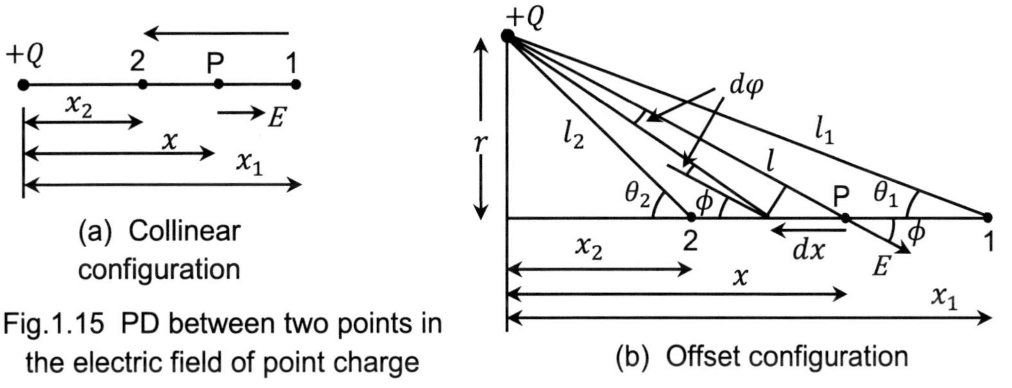

Fig.1.15 PD between two points in the electric field of point charge

(b) Offset configuration: Fig.1.15(b) illustrates this situation. Field at point P due to the charge $+Q$ is E, but only its horizontal component $E\,Cos\,\phi$ contributes in moving a charge along the horizontal straight line passing through the points 1 and 2. Hence, we have $V_{21} = -\int_1^2 E\,Cos\,\phi\,.dx = -\frac{Q}{4\pi\varepsilon}\int_1^2 \frac{Cos\,\phi}{l^2}.dx$. In line with the mathematics used in Sec.1.7, here we have $dx.Sin\,\phi = l.d\varphi$ and $Sin\,\phi = r/l$. Therefore, $dx\frac{r}{l} = l.d\varphi$, or, $\frac{dx}{l^2} = \frac{d\varphi}{r}$. Further, as x decreases by dx, ϕ increases by $d\varphi$, and this necessitates a minus sign for integration, $(-)\frac{dx}{l^2} = \frac{d\varphi}{r}$. As such,

$$V_{21} = \frac{Q}{4\pi\varepsilon}\int_{\theta_1}^{\theta_2}\frac{Cos\,\phi}{r}.d\varphi = \frac{Q}{4\pi\varepsilon r}\int_{\theta_1}^{\theta_2} Cos\,\phi\,.d\varphi = \frac{Q}{4\pi\varepsilon r}[\,Sin\,\phi\,]_{\theta_1}^{\theta_2}$$

$$= \frac{Q}{4\pi\varepsilon r}\;(Sin\,\theta_2 - Sin\,\theta_1) = \frac{Q}{4\pi\varepsilon r}\left(\frac{r}{l_2} - \frac{r}{l_2}\right) = \frac{Q}{4\pi\varepsilon l_2} - \frac{Q}{4\pi\varepsilon l_1} = V_2 - V_1$$

The obvious result is the same as in the collinear case, but the deriving principle in this case is of conceptual significance, and will be applied in some later derivations.

1.8.3 Potential of charged spherical conductor: By saying 'potential of a conductor', we actually mean the potential at points on the surface of the conductor. Field strength at any point on the surface of a spherical conductor of radius R and carrying a charge Q being $E = Q/(4\pi\varepsilon R^2)$, it follows from Eqn.1.21 that the potential (at any point on the surface) of a spherical conductor of radius R and carrying a charge Q (with respect to zero reference at infinite distance) is given by

$$V = \frac{Q}{4\pi\varepsilon R} \qquad \text{...} (1.22)$$

where ε is the permittivity of the medium outside the sphere.

If all the points on any line or surface in an electric field have the same or uniform value of potential, it is called an *equipotential line* or *equipotential surface*. For example, the surface of the imaginary sphere in Fig.1.4 is an equipotential surface. In fact, the entire surface of a charged conductor *of any geometry* is an equipotential surface. Hence, the field strength or potential gradient along any equipotential line or surface is zero. At any point on the surface of a charged conductor, the field strength or the flux density is normal to the surface at that point with no component along the surface. This is obvious, because had there been any component along the surface, charge flow would occur, the surface being conducting. It therefore follows that although the charge density at points on the surface of a charged conductor (other than a sphere) varies with the surface curvature, the potential remains the same.

In terms of the PE_E described in Sec.1.8, the 'PE_E of a charged conductor' (far distant from any other charge) is an invalid statement (whereas the 'SE_E of a charged conductor', Sec.1.14, is quite meaningful).

1.8.4 Potential and potential energy in the field of single charged plate: In Fig.1.16(a), AB is a thin circular conducting plate of radius R, charged to surface density $+\sigma$, and P is a point distant x from the centre of the plate on its axis. Field strength at P was derived in Sec.1.7.2(a) as $\frac{\sigma}{2\varepsilon}(1 - Cos\theta)$. At $x = 0,\ \theta = \pi/2,\ Cos\theta = 0$ and $E = \frac{\sigma}{2\varepsilon}$.

The E-x variation is as shown by curve 2-3-1 in Fig.1.16(b). The same curve represents, to different scale, the force F on some positive charge at P. Potential at point P

$$V = -\int_{x=\infty}^{x=x} E.dx = -\frac{\sigma}{2\varepsilon}\int_{x=\infty}^{x=x}(1 - Cos\theta)dx.$$

Length CD $= l.d\theta = dx.Sin\theta$ and $Sin\theta = R/l$. So, $\frac{R}{Sin\theta}d\theta = dx.Sin\theta$, or $dx = \frac{R}{Sin^2\theta}d\theta$.

As x decreases by dx, θ increases by $d\theta$. Hence, $-dx = \frac{R}{Sin^2\theta}d\theta$ and

$$-\frac{\sigma}{2\varepsilon}(1 - Cos\theta)dx = \frac{\sigma R}{2\varepsilon}.\frac{1-Cos\theta}{Sin^2\theta}d\theta = \frac{\sigma R}{2\varepsilon}.\frac{d\theta}{1+Cos\theta}.$$

Thus,
$$V = \frac{\sigma R}{2\varepsilon}\int_0^{\theta}\frac{d\theta}{1+Cos\theta} = \frac{\sigma R}{2\varepsilon}\left[tan\frac{\theta}{2}\right]_0^{\theta}$$

$$= \frac{\sigma R}{2\varepsilon}tan\frac{\theta}{2} \qquad \text{.................................} (1.23(a))$$

The V-x variation is, therefore, given by the $tan\,\theta/2$ curve 4-5-1 in Fig.1.16(b) within the range from $\theta/2 = \pi/4$ at $x = 0$ to $\theta/2 = 0$ at $x = \infty$.

By the identity $tan\frac{\theta}{2} = \sqrt{\frac{1-Cos\theta}{1+Cos\theta}} = \frac{Sin\theta}{1+Cos\theta} = \frac{1-Cos\theta}{Sin\theta}$ we have,

$$tan\frac{\theta}{2} = \frac{1-Cos\theta}{Sin\theta} = \frac{1-x/l}{R/l} = \frac{l-x}{R} = \frac{1}{R}\left[\sqrt{x^2+R^2} - x\right]$$

As such, we may also express the potential as

$$V = \frac{\sigma}{2\varepsilon}\left[\sqrt{x^2+R^2} - x\right] \qquad \text{.. (1.23)}$$

This potential is with respect to zero reference at $x = \infty$. Curve 08 is the plot the potential with respect to zero reference at $x = 0$, that is, with respect to the plate AB. Thus, the potential V of the point P with respect to zero reference at $x = \infty$ is the height 75, which is positive, and that with respect to the plate AB is the depth 46 or 78, which is negative. The same potential curves also represent, to different scale, the PE_E of any positive charge at the point P.

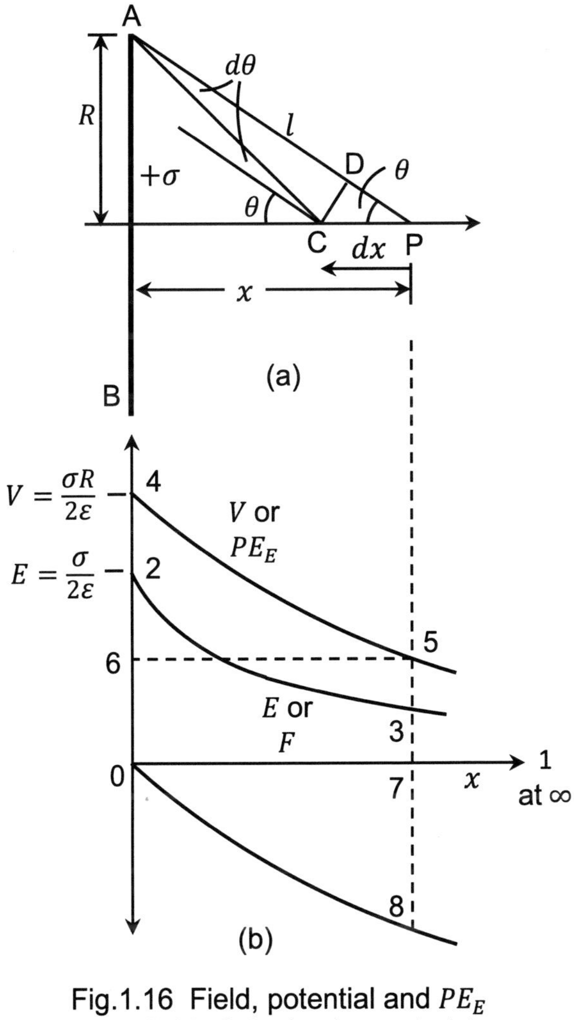

Fig.1.16 Field, potential and PE_E in the field of single charged plate

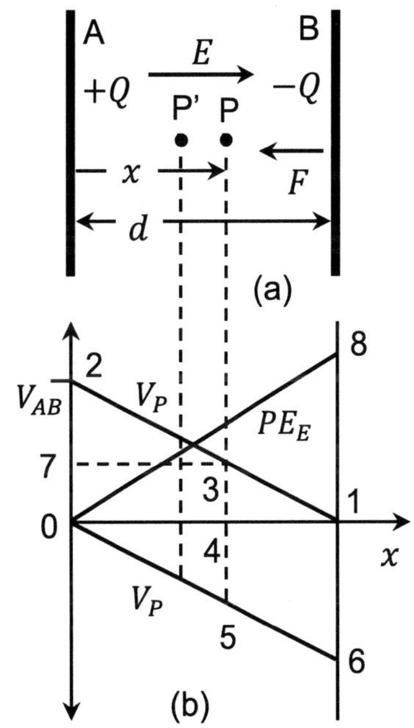

Fig.1.17 Potential, PD and PE_E in the field of charged parallel plates

1.8.5 Potential, PD and PE in the field of charged parallel plates: In Fig.1.17(a), the field(force on a test charge) at any point P between two close parallel conducting plates with charges $\pm Q$, neglecting fringing effect, was derived in Sec.1.7.2, as $E = Q/(\varepsilon A)$, directing from the positive plate A to the negative plate B. Hence, work done by the field in moving a test charge from point P to plate B, that gives the potential of point P with respect to that of plate B, $V_P = E(d-x) = \frac{Q}{\varepsilon A}(d-x)$. As E is constant for given values of $\pm Q$, V_P-x relation is linear. Here the field does not extend upto infinity but terminates on plate B ($x = d$). Hence, the zero reference of potential for points like P is the position of plate B. Against this zero reference, line 12 in Fig.1.17(b) is the plot of V_P along the distance x. Potential of point P with respect to that of plate A, $V_P = -Ex = -\frac{Q}{\varepsilon A}x$. Line 06 is the plot of V_P with respect to zero reference at plate A ($x = 0$). Thus, the potential V_P of the point P with respect to zero reference at plate B, height 43, is positive, and that with respect to zero reference at plate A, depth 27 or 45, is negative. Considering another point like P', it can, however, be seen that the PD between the two points remains unaffected by the selection of zero reference, as was mentioned earlier in Sec.1.8. By the term 'PD between the plates', we mean

$$V = V_{AB} = Ed = \frac{Q}{\varepsilon A}d \qquad \text{.................................... (1.24)}$$

The uniform field at all points on plate B due to charge $+Q$ on plate A alone is $E_A = Q/(2\varepsilon A)$. Plate B that carries a charge $-Q$ and lies in field E_A experiences an attractive (negative) force $F = E_A.Q = [Q/(2\varepsilon A)]\,Q = Q^2/(2\varepsilon A)$. Thus, like E, F is also constant for given value of $\pm Q$.

We now consider the PE_E of the plates. Potential at any point on, say, plate B, with zero reference at and due to charge $+Q$ on plate A alone, is $-\frac{Q}{2\varepsilon A}d$, and plate B, with charge $-Q$, lies at this potential. So, the PE_E of plate B in the field of and with zero reference at plate A, $(PE_E)_{BA} = \left(-\frac{Q}{2\varepsilon A}d\right)(-Q) = \frac{Q^2}{2\varepsilon A}d = Fd$. As F is constant, using x for the distance d between the plates, the $(PE_E)_{BA}$ - x relationship is also linear, line 08. As x increases, PE_E of plate B increases, so dPE_E/dx is positive. By writing $F = -dPE_E/dx$ or $-F = d(PE_E)_{BA}/dx$, we merely signify that the direction of F on plate B is negative (F acts along negative direction of x). This implies that the amount of mechanical energy expended (Fx) by some external agency in pulling apart plate B through distance x from plate A against attraction F is stored as positive PE_E of plate B in the field of and with zero reference at plate A. We may also write that

$$(PE_E)_{BA} = \left(-\frac{Q}{2\varepsilon A}d\right)(-Q) = [(-V/2)(-Q)] = +VQ/2$$

The same will be the PE_E of plate A in the field of and with zero reference at plate B

$$(PE_E)_{AB} = [(+V/2)(+Q)] = +VQ/2.$$

Hence, in the attractive electric field between two charged conductors, the PE_E of either conductor in the field of, and with zero reference at, the other conductor is $VQ/2$, and is positive. This interpretation will be found analogous to the magnetic case in Sec.4.17.1.

1.8.6 Potential near short electric dipole: In Sec.1.7.5, we derived the field strength near a short electric dipole. Derivation for potential is rather simpler because potential does not bear direction. Thus, reproducing Fig.1.11(b) as Fig.1.18 here, we derive the potential of point P distant x from the centre of the dipole as

$$V = \frac{+Q}{4\pi\varepsilon.(\mathrm{PB})} + \frac{-Q}{4\pi\varepsilon.(\mathrm{PA})}$$

As was derived in Sec.1.7.5, when $x \gg d$,

PB $= x - (d/2)Cos\theta$, and, PA $= x + (d/2)Cos\theta$.

Putting these values of PB and PA, we get

$$V = \frac{Q}{4\pi\varepsilon}\left[\frac{1}{x-(d/2)Cos\theta} - \frac{1}{x+(d/2)Cos\theta}\right]$$

$$= \frac{Q}{4\pi\varepsilon}\left[\frac{(x+(d/2)Cos\theta)-(x-(d/2)Cos\theta)}{x^2-\{(d/2)Cos\,\theta\}^2}\right]$$

$$= \frac{Q}{4\pi\varepsilon}\frac{d\,Cos\theta}{[x^2-\{(d/2)Cos\,\theta\}^2]}$$

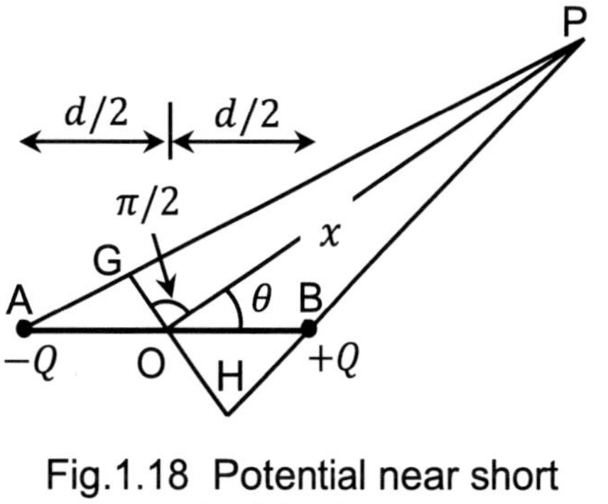

Fig.1.18 Potential near short electric dipole

As $x \gg d$, the denominator within the square bracket reduces to x^2, and

$$V = \frac{QdCos\theta}{4\pi\varepsilon x^2} \quad \text{.................................} \quad (1.25)$$

This potential is with respect to zero reference at infinite distance. On the axis of the dipole, $\theta = 0$, $V = Qd/(4\pi\varepsilon x^2)$. On the perpendicular bisector plane of the dipole, $\theta = \pi/2$, $V = 0$. The perpendicular bisector plane is thus an equipotential surface of zero potential. It is noticed that potential near dipole varies as $(1/x^2)$, while that near a point charge varies as $(1/x)$.

1.9 Electric Dipole Moment: Dipole moment of an electric dipole constituted by two equal and opposite point charges $\pm Q$ separated by a distance d, specifies the strength of the dipole. It is defined as a vector denoted by

$$\boldsymbol{p_e} = Q.\boldsymbol{d} \quad \text{...} \quad (1.26)$$

directed from $-Q$ to $+Q$, as shown at Fig.1.19(a).

The subscript e denotes 'electric'. Significance of this definition can be appreciated from the following consideration. Dipole moment being a vector, its magnitude and direction parts need separate considerations.

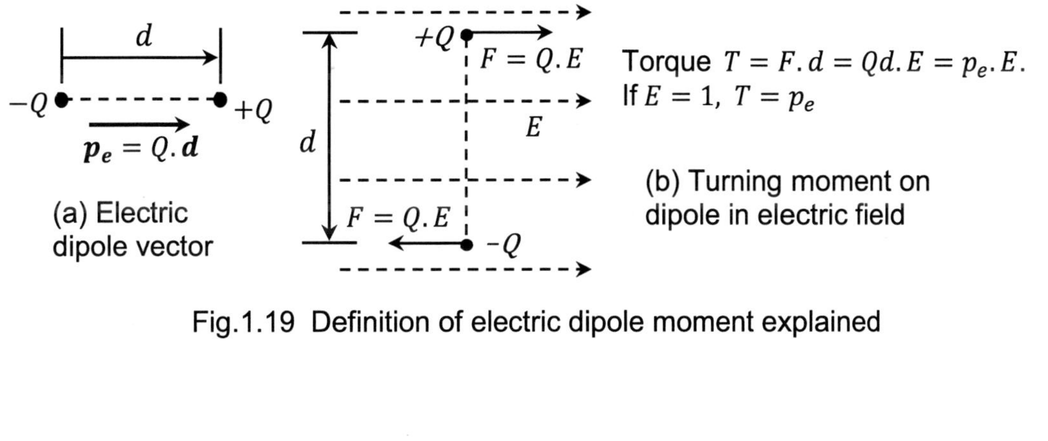

Fig.1.19 Definition of electric dipole moment explained

Magnitude Part: The electric effect of any system of charges manifests as electric field strength and electric potential *surrounding* the system. In Sec.1.7.5 it was seen that the field strength near an electric dipole is proportional to the product $Q.d$. It was also seen in Sec.1.8.6 that the potential near an electric dipole is proportional to the product $Q.d$. These observations reveal that the strength or magnitude of an electric dipole is the product $Q.d$.

Direction Part: If a number of neighbouring electric dipoles of different magnitudes lie with their axes aligned in different three dimensional orientations in space, what would be the equivalent dipole moment of the system? The question can be answered only if some direction be assigned to the dipole moment, because, by assigning such a direction, the vector sum of the neighbouring dipole moments will give the resultant dipole moment. Although any direction could be chosen arbitrarily, the most logical and conventional one is that of the dipole field along the axis of the dipole and *outside* it; 'outside' because we are concerned with the electric effect of the dipole at points outside the dipole. Hence the direction of a dipole is defined from the negative charge $-Q$ to the positive charge $+Q$, as shown in Fig.1.19(a), and as such, the distance $\boldsymbol{d}$ is treated as a vector.

In the term Dipole Moment, the word 'moment' owes from the following fact. If a dipole is placed in an electric field of strength E, with its axis normal to E, as shown in Fig.1.19(b), the two equal and opposite parallel forces $F = Q.E$ acting on the two charges of the dipole form a couple, and the turning moment or torque of the couple is given by $T = F.d = Qd.E = p_e.E$. If $E = 1$, then $T = p_e = Qd$. The concept of electric dipole moment is of particular importance in the atomic study of dielectric materials.

1.10 Capacitance: Capacitance is the charge storing capacity of a conductor or a system of conductors. The earlier term '*capacity*' is now obsolete. It is denoted by C. Eqn.1.22 shows that the potential of a charged conducting sphere is proportional to its charge. In fact, this proportionality is universal to all geometries of the conductor.

Capacitance of a conductor is defined as the quantity of positive electric charge that it requires to raise its electric potential by one unit.

Thus, in general, if a charge Q raises a conductor to a potential V, its capacitance is given by

$$C = Q/V \qquad \text{...} \quad (1.27)$$

It follows from Eqn.1.22 that the capacitance of a spherical conductor, in terms of its geometry,

$$C = Q/V = 4\pi\varepsilon R \qquad \text{...} \quad (1.28)$$

This equation leads us to the logical conclusion that the capacitance of a conductor is proportional to its size or dimension. This is obvious, because the larger a conductor more charge is required to raise its potential by a given amount, and is analogous to the fact that the larger a tank more water is required to raise its water level by a given height. The capacitance is thus a measure of how much charge can be stored with how less potential. The device for storage of electric charge is called a capacitor. The earlier term '*condenser*' is now obsolete. A capacitor comprises of two conductors separated by some dielectric medium. The two parallel conducting plates of Fig.1.6 form a *parallel plate capacitor*. Practical capacitors assume various shapes and sizes depending on the geometry and spacing of the conductors.

If a capacitor be connected to a DC source, say a storage battery, Fig.1.20, it is charged with the PD of the source between its two conductors, and equal quantities of positive and negative charges appear on its two conductors.

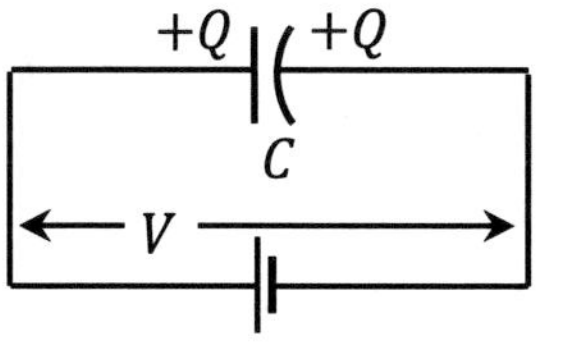

Fig.1.20 Capacitor explained

Capacitance of a capacitor is defined as the quantity of equal and opposite electric charges on its two conductors per unit electric potential difference between them.

It follows from Eqn.1.24 that the capacitance of a parallel plate capacitor, in terms of its geometry, is given by

$$C = \frac{Q}{V} = \frac{\varepsilon A}{d} \quad \text{.. (1.29)}$$

This equation shows that the capacitance of a capacitor is directly proportional to its plate area and inversely proportional to its dielectric thickness. Unit of capacitance is farad (F), which is coulomb/volt (C/V). Both the Eqns.1.28 and 1.29 reveal that the capacitance is

(a) directly proportional to the permittivity of the dielectric, and

(b) it is dimensionally (permittivity x length), which explains the unit of permittivity as farad per metre.

1.10.1 Capacitors in series: Fig.1.21 shows three capacitors C_1, C_2 and C_3 connected in series across a DC source of V volt. The charge $+Q$ appearing on the positive plate of the first capacitor induces an equal and opposite charge $-Q$ on its negative plate. This in turn leaves a charge $+Q$ on the positive plate of the next capacitor, and so on. Hence, the charge on all the capacitors is the same.

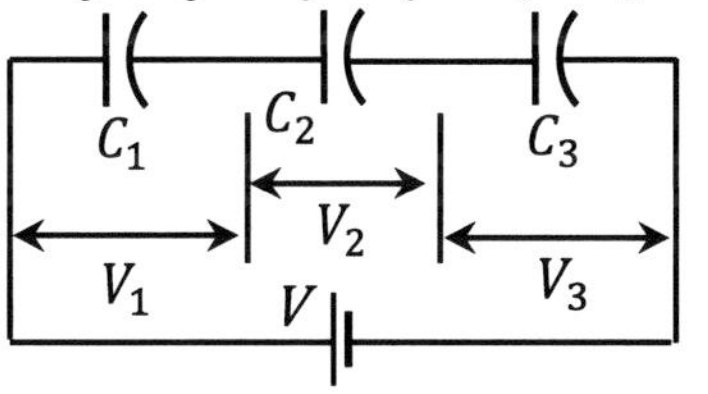

Fig.1.21 Capacitors in series

But their capacitances being different, voltages across them are different, satisfying the relation $V = Q/C$, the total voltage being the source voltage V. Thus, in general, for n number of capacitors in series we have

$$V = V_1 + V_2 + V_3 + \cdots V_n = \frac{Q}{C_1} + \frac{Q}{C_2} + \frac{Q}{C_3} + \cdots \frac{Q}{C_n}$$

$$= Q\left[\frac{1}{C_1} + \frac{1}{C_2} + \frac{1}{C_3} + \cdots \frac{1}{C_n}\right]$$

$$= Q\left[\frac{1}{C}\right],$$

where C is the single capacitance equivalent to $C_1, C_2, C_3, \ldots . C_n$ in series. Hence, C is given by

$$\frac{1}{C} = \frac{1}{C_1} + \frac{1}{C_2} + \frac{1}{C_3} + \cdots \frac{1}{C_n} \quad \text{.................................... (1.30)}$$

This shows that adding capacitors in series reduces the equivalent capacitance.

1.10.2 Capacitors in parallel: Fig.1.22 shows three capacitors C_1, C_2 and C_3 connected in parallel across a DC source of V volt. The capacitors draw charges $\pm Q_1$, $\pm Q_2$ and $\pm Q_3$, as is given by the relation $Q = C.V$. Thus, in general, for n number of capacitors in parallel, the total charge drawn from the source

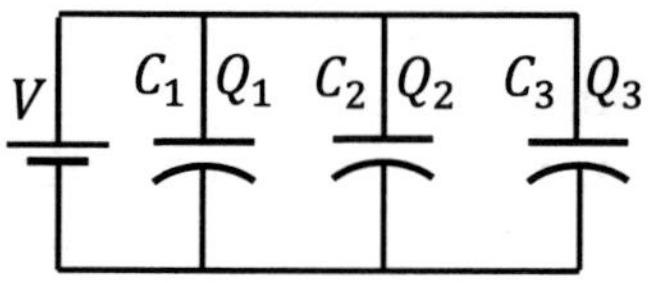

Fig.1.22
Capacitors in parallel

$$Q = Q_1 + Q_2 + Q_3 + \cdots Q_n$$
$$= C_1.V + C_2.V + C_3.V + \cdots C_n.V$$
$$= V(C_1 + C_2 + C_3 + \cdots C_n) = V.C$$

where C is the single capacitance equivalent to $C_1, C_2, C_3, \ldots. C_n$ in parallel. Hence,

$$C = C_1 + C_2 + C_3 + \cdots C_n \quad \ldots\ldots\ldots\ldots (1.31)$$

This shows that adding capacitors in parallel increases the equivalent capacitance. These results are just opposite to the corresponding cases of resistances in series and in parallel (Current Electricity).

1.10.3 Capacitors with composite dielectric:

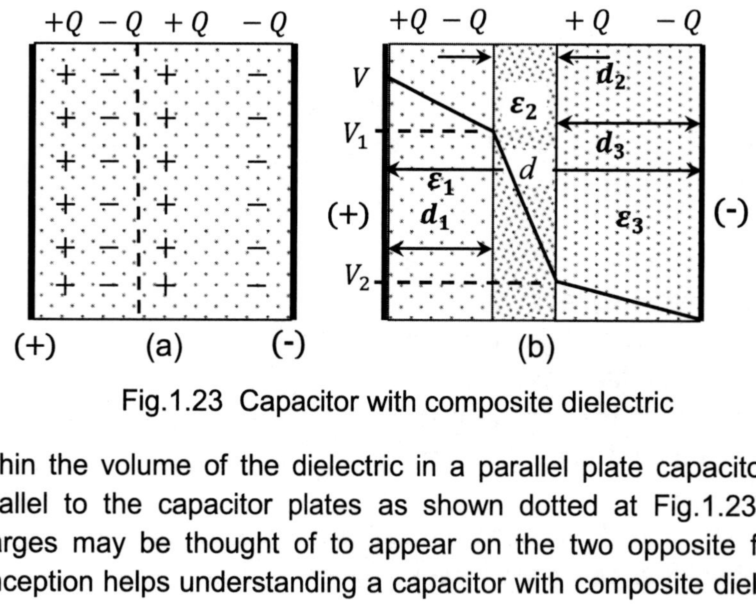

Fig.1.23 Capacitor with composite dielectric

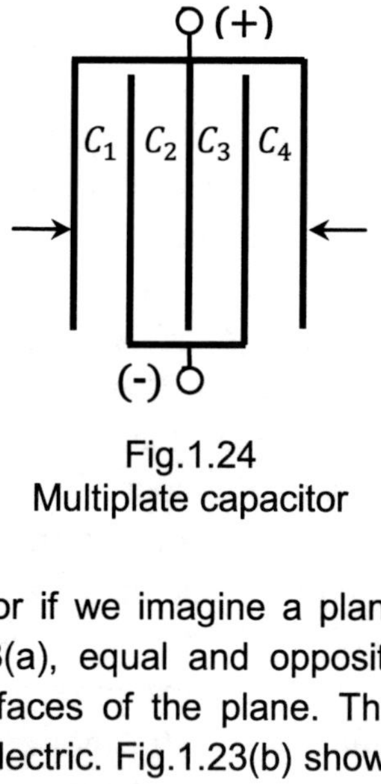

Fig.1.24
Multiplate capacitor

Within the volume of the dielectric in a parallel plate capacitor if we imagine a plane parallel to the capacitor plates as shown dotted at Fig.1.23(a), equal and opposite charges may be thought of to appear on the two opposite faces of the plane. This conception helps understanding a capacitor with composite dielectric. Fig.1.23(b) shows a parallel plate capacitor with three different dielectrics of permittivity ε_1, ε_2, ε_3 and thickness d_1, d_2, d_3 respectively. Equal and opposite charges appear on the two opposite faces of the planes at the junction of two adjacent dielectric layers, and this, in effect, forms three capacitors in series. As such, the combined capacitance C of a capacitor with plate area A and with multiple dielectrics is given by Eqn.1.30 as

$$\frac{1}{C} = \frac{1}{C_1} + \frac{1}{C_2} + \frac{1}{C_3} + \cdots \frac{1}{C_n} = \frac{d_1}{\varepsilon_1 A} + \frac{d_2}{\varepsilon_2 A} + \frac{d_3}{\varepsilon_3 A} + \cdots \frac{d_n}{\varepsilon_n A}$$

$$\text{or, } C = \frac{A}{\frac{d_1}{\varepsilon_1} + \frac{d_2}{\varepsilon_2} + \frac{d_3}{\varepsilon_3} + \cdots \frac{d_n}{\varepsilon_n}} \quad \ldots\ldots\ldots\ldots (1.32)$$

The electric flux density D in all the dielectric layers is the same but the potential gradient or the electric field strength $E\ (= D/\varepsilon)$ varies inversely with the permittivity of the dielectric. Hence, the PD across the dielectric layers $(V = E.d)$ may be different. Typical shape of potential distribution is graphically represented in Fig.1.23(b). In electrical power engineering, one application of composite dielectric is 'capacitance grading' of high voltage cables.

1.10.4 Multi-plate capacitors: Fig.1.24 shows a capacitor with five plates, three positive and two negative. The positive and negative sets are inserted one inside other, as shown. With this arrangement, both surfaces of all the inner plates are utilised, thereby increasing the effective surface areas of the plates. Out of the five plates, eight surfaces are utilised, forming in effect four capacitances C_1, C_2, C_3 and C_4 in parallel across the positive and negative terminals. The equivalent capacitance is then given by Eqn.1.31 as $C = C_1 + C_2 + C_3 + C_4$. If the plate areas A, inter-plate thicknesses d and the inter-plate dielectrics are the same, then $C_1 = C_2 = C_3 = C_4 = \varepsilon A/d$. As n number of plates form $(n - 1)$ number of capacitances in parallel, we have

$$C = (n-1)\,\frac{\varepsilon A}{d} \qquad \text{...} \quad (1.33)$$

This arrangement reduces the size of the capacitor for some given value of capacitance.

1.10.5 Self capacitance mutual capacitance and earth capacitance:

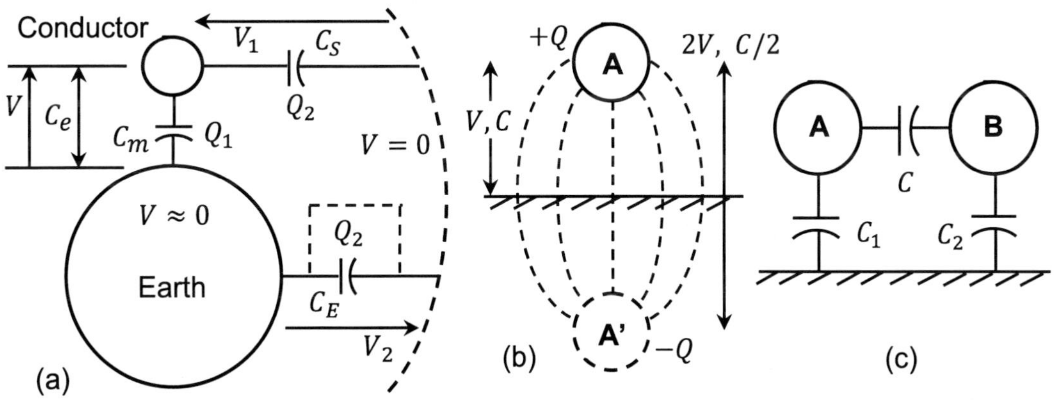

Fig.1.25 Earth capacitance explained

The capacitance of a conductor, when it lies in isolation, far remote from any other conductor including the earth, which latter is also a conductor of huge size, is called its *self capacitance*. The earth as a conductor has also its own self capacitance. On the other hand, the capacitance of a capacitor is the *mutual capacitance* between its two conductors. In fact, a capacitance can exist between two conductors only. As the self capacitance of a conductor is given by its (charge/potential) ratio, and the potential is reckoned from its zero reference at infinite distance, the self capacitance of a conductor is thought of as its mutual capacitance with an imaginary enveloping hollow conductor of infinite size centred at the conductor, shown dotted in Fig.1.25(a). The value of self capacitance depends solely on the shape and size of the conductor. Self capacitance of a spherical conductor of radius R being $4\pi\varepsilon_0 R$, its typical value for, say, a 20-cm radius conducting ball works out as C_s = 22 pF or 22x10^{-6} μF. Against this, the self capacitance of the earth (radius 6,371 km), C_E = 710 μF; although enormously high compared to C_s , it is significantly smaller than a farad. For a conductor on the earth, therefore, three capacitances come into effect; these are C_s, C_E and the mutual capacitance between the conductor and the earth, C_m. These are shown in Fig.1.25(a). The effective or resultant capacitance between a conductor and the earth is defined as the *earth capacitance* C_e of the conductor. From Fig.1.25(a), $C_e = C_m + \frac{C_S.C_E}{C_S+C_E}$. As $C_E \gg C_s$, we have $C_e \approx C_m + C_s$. Whereas C_s depends only on the shape and size of the conductor, C_m depends on additional factors like its distance from, and its orientation with respect to, the earth. Theoretically, as C_m tends to zero, C_e tends to C_s. In practice, any conductor being too closer to the earth than to the imaginary hollow conductor of infinite size, C_m is much greater than C_s , and so the earth capacitance of a conductor is practically given by its mutual capacitance to earth. If a charge Q is imparted to the conductor, it divides into two parallel circuits to the earth, Q_1 in C_m, and Q_2 in C_s and C_E in series. As $C_m \gg \frac{C_S.C_E}{C_S+C_E}$, almost all the charge is drawn by C_m, that is $Q_1 \gg Q_2$. The PDs across the three capacitors are shown in the figure. In the series circuit of C_s and C_E, we have $V = V_1 + V_2$ and $Q_2 = C_S.V_1 = C_E.V_2$. Hence, $\frac{V_1}{V_2} = \frac{C_E}{C_S}$. Since $C_E \gg C_S$, $V_1 \gg V_2$. Hence, $V_1 \approx V$ and $V_2 \approx 0$, as if the capacitor C_E is shorted, as indicated in the figure. This means, the potential of the earth, with respect to the zero reference of the imaginary hollow conductor of infinite size, is also zero for all practical purposes. This is the logic the earth is considered to have a zero potential. The enveloping hollow conductor of infinite size being imaginary, V_1 cannot be measured, and so the self capacitance C_s is only of theoretical significance. On the other hand, V is a measurable realistic quantity. For all conductors on the earth, therefore, it is the earth capacitance C_e that has to be considered, rather than the self capacitance C_s , and C_e is again given by the mutual capacitance to earth C_m.

When a charged conductor lies above the earth, Fig.1.25(b), the lines of force between the conductor and the earth are so profiled as if an image of the conductor is there below the mirror-acting earth surface. This means, the PD and mutual capacitance between the conductor and its image are respectively $2V$ and $C/2$, where V is the potential of the conductor above the zero-potential earth surface and C its earth capacitance. This concept of *electrical image* helps deriving the earth capacitance of practical conductors.

When two or more conductors lie in proximity, like the two conductors of a capacitor, they possess their individual earth capacitances as well as mutual capacitances between them. As such, two neighbouring conductors A and B (above the earth), Fig.1.25(c), remote from any other conductor, are associated with three capacitances, their earth capacitances C_1 and C_2, and their mutual capacitance C. The conductors may be of any shape, not necessarily the spherical ones shown in the figures.

In general, in a neighbouring multiple-conductor system, any charge imparted to any one conductor induces different charges on all the other conductors depending on all the earth and mutual capacitances, and thereby modifies the charges and potentials of all the conductors. This forms the basis of the *generalised theory of potential and capacitance* developed in terms of some *potential and capacitance coefficients*, and expressed as a set of linear simultaneous equations. For better appreciation of the subject, we develop below an alternative concept for two conductors. Fig.1.26 illustrates this concept.

In Fig.1.26(a), the conductor A lies far remote from any other conductor like B. A charge $+Q_1$ imparted to the conductor raises its potential (with respect to the zero reference of the earth) to $V_1' = Q_1/C_1$. This is shown graphically at Fig.(e). The uncharged conductor B is then brought close to conductor A, so that a mutual capacitance C occurs between them, Fig.(b). Now the charge Q_1 on conductor A sees two parallel paths to the earth, one through C and C_2 in series, and another through C_1. The charge Q_1, therefore, divides accordingly in two parts q_{11} on C_1, and q_A on both C and C_2, as shown in the figure. The charge q_A gets engaged in C, so does not contribute to the potential of A with respect to earth, which latter, therefore, owes to the charge q_{11} only. As such, the potential of A drops from $V_1' = Q_1/C_1$ to $v_{11} = q_{11}/C_1$, as shown at Fig.(e). Although no charge has so far been imparted to conductor B, it gains an induced charge q_A and an induced potential $v_{21} = q_A/C_2$ with respect to earth, by induction from A. If a person standing on the earth happens to touch conductor B, he will get electric shock of the potential v_{21}. Similarly, when conductor B is imparted a charge $+Q_2$, with no charge on A, Fig.(c), the two parallel paths for Q_2 to earth are through C and C_1 in series, and through C_2. In this case, Q_2 divides in two parts q_{22} on C_2, and q_B on both C and C_1. Accordingly, the potentials of A and B are respectively $v_{12} = q_B/C_1$ and $v_{22} = q_{22}/C_2$. Finally, if both the charges $+Q_1$ and $+Q_2$ be there on A and B simultaneously, Fig.(d), the total charges on C_1, C_2 and C become, by superimposition, $(q_{11} + q_B)$, $(q_{22} + q_A)$ and $(q_A - q_B)$ respectively. It can be seen that the total charge on any conductor does not change by the proximity of, or charge on, a neighbouring conductor, only its distribution to earth and mutual capacitances modifies. For example, as $+Q_2$ is imparted to B, the charge on C_1 increases from q_{11} to $(q_{11} + q_B)$ and that on C reduces from q_A to $(q_A - q_B)$, the total charge on conductor A, $Q_1(= q_{11} + q_A)$ remaining the same. Final potentials of A and B due to charges $+Q_1$ and $+Q_2$ on them are respectively $V_1 = v_{11} + v_{12}$ and $V_2 = v_{22} + v_{21}$, and their PD $V_{AB} = V = V_1 - V_2$. The potentials of both the conductors thus increase, but their PD decreases. In Fig.(e), line 2 is less inclined than line 1, and this represents the decrease in the PD. As we also have

$$q_A = C(v_{11} - v_{21}) = C_2 v_{21} \text{ and}$$
$$q_B = C(v_{22} - v_{12}) = C_1 v_{12}$$

we can write for Q_1 and Q_2 as

$$Q_1 = q_{11} + q_A = (q_{11} + q_B) + (q_A - q_B)$$
$$= (C_1 v_{11} + C_1 v_{12}) + [C(v_{11} - v_{21}) - C(v_{22} - v_{12})]$$
$$= C_1(v_{11} + v_{12}) + C(v_{11} + v_{12}) - C(v_{22} + v_{21})$$
$$= C_1 V_1 + C(V_1 - V_2)$$

Similarly, $Q_2 = C_2 V_2 + C(V_2 - V_1) = C_2 V_2 - C(V_1 - V_2)$

Writing these relations as a set,

$$\left.\begin{aligned} Q_1 &= C_1 V_1 + C(V_1 - V_2) = C_1 V_1 + CV \\ Q_2 &= C_2 V_2 - C(V_1 - V_2) = C_2 V_2 - CV \end{aligned}\right\} \quad \text{................ (1.34)}$$

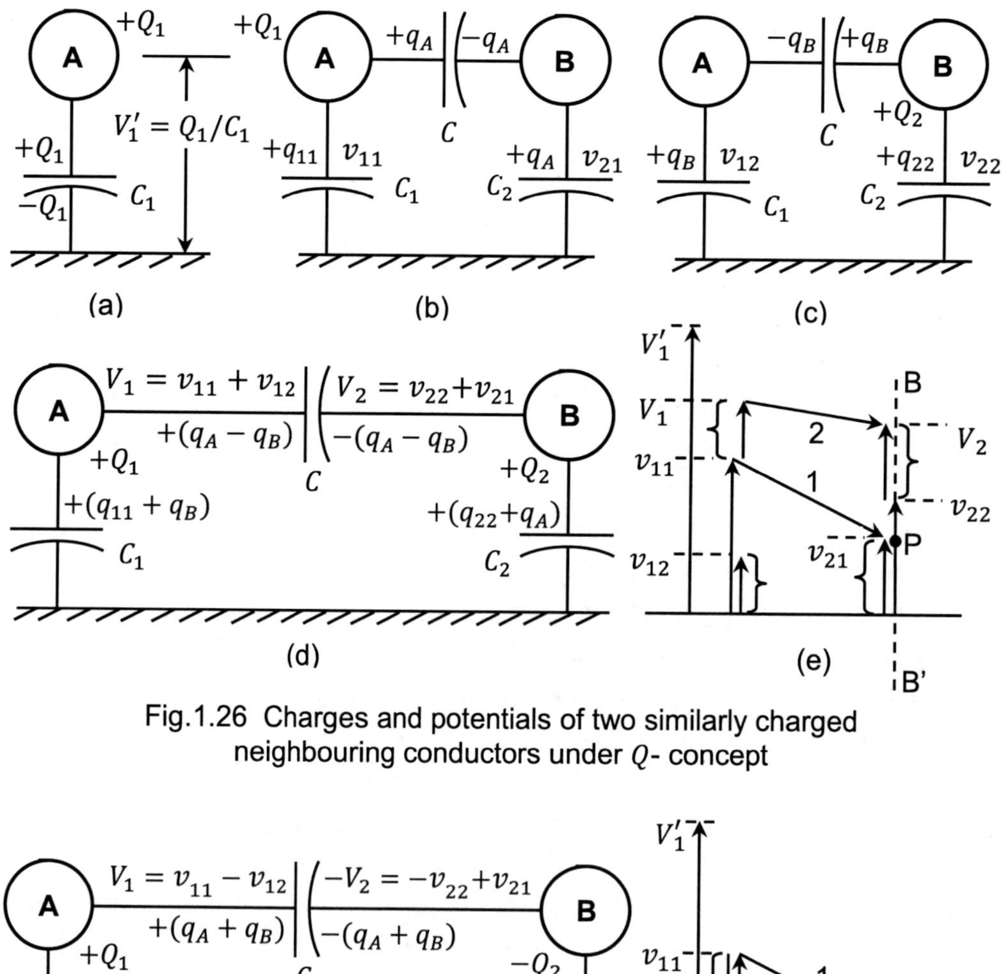

Fig.1.26 Charges and potentials of two similarly charged neighbouring conductors under Q- concept

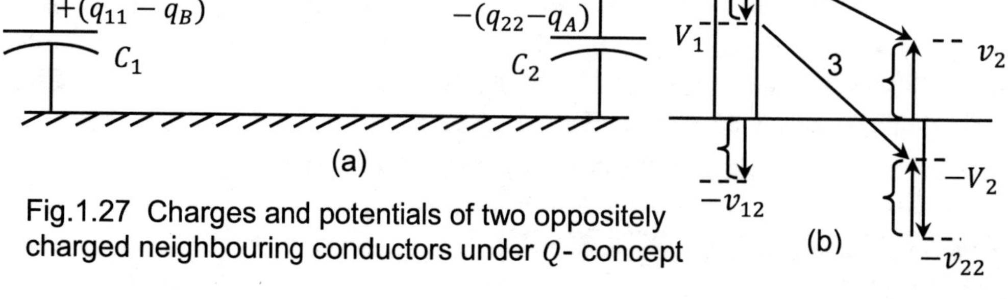

Fig.1.27 Charges and potentials of two oppositely charged neighbouring conductors under Q- concept

These relations simply assert that the terms C_1V_1, C_2V_2 and $C(V_1 - V_2)$ are respectively the charges on C_1, C_2 and C. Eqns.1.34 relate the charges potentials and PD of the two conductors. Let us study how the potentials and PD vary as the charge on any one conductor, say, Q_2 on conductor B, varies, when the charge $+Q_1$ on the other conductor A is constant. With this condition, Eqns.1.34 may assume different forms under different situations. When $Q_2 = 0$, all the V_1, V_2 and V are positive. $V_2(= v_{21})$ is given by the point P in Fig.1.26(e).

A given set of values of C_1, C_2, C, and Q_1, therefore, locates the position of the point P on the line BB'. As the positive charge $+Q_2$ increases from zero, V_1 increases, V_2 rises above the point P, and V decreases from its value given by line 1.

When $+Q_2$ increases to the proportion $(Q_1/Q_2) = (C_1/C_2)$, it can be derived that $V_1 = V_2$, and so V becomes zero. This derivation follows from Eqns.1.34

$$\left.\begin{aligned} Q_1C_2 &= C_1C_2V_1 + CC_2V_1 - CC_2V_2 \\ Q_2C_1 &= C_1C_2V_2 - CC_1V_1 + CC_1V_2 \end{aligned}\right\}$$

If $(Q_1/Q_2) = (C_1/C_2)$, or $Q_1C_2 = Q_2C_1$, we have the identity

$$C_1C_2V_1 + CC_2V_1 - CC_2V_2 = C_1C_2V_2 - CC_1V_1 + CC_1V_2$$

or, $$V_1(C_1C_2 + CC_2 + CC_1) = V_2(C_1C_2 + CC_1 + CC_2)$$

or, $V_1 = V_2$

This equipotential situation (maintaining C, C_1 and C_2 unaltered) can be created in two possible ways; (a) by bringing the two conductors into contact, and then separating them back to their original positions, or (b) by connecting them by a thin wire. If $+Q_2$ is greater than that given by this equipotential condition, V_2 will be greater than V_1, giving $V = V_2 - V_1$; under this situation, Eqns.1.34 assume the form

$$\left.\begin{aligned} Q_1 &= C_1V_1 - C(V_2 - V_1) = C_1V_1 - CV \\ Q_2 &= C_2V_2 + C(V_2 - V_1) = C_2V_2 + CV \end{aligned}\right\} \qquad \text{................. (1.34(a))}$$

As Q_2 decreases from zero, or becomes negative $(-Q_2)$, V_1 decreases, V_2 lowers below the point P, and V increases from its value given by line 1 in Fig.1.26(e). For small values of $-Q_2$, the net charge on C_2, and so V_2 also, remain positive. In other words, in the second line of Eqns.1.34, the negative term CV becomes larger than the positive term C_2V_2, making Q_2 negative. In this zone, Eqns.1.34 assume the form

$$\left.\begin{aligned} Q_1 &= C_1V_1 + C(V_1 - V_2) = C_1V_1 + CV \\ -Q_2 &= C_2V_2 - C(V_1 - V_2) = C_2V_2 - CV \end{aligned}\right\} \qquad \text{................. (1.34(b))}$$

V_2 becomes zero $(V = V_1)$ with some negative charge $Q_2 = CV = CV_1 = CQ_1/(C_1 + C)$. As Q_2 becomes more negative, V_2 becomes negative, giving $V = V_1 - (-V_2) = V_1 + V_2$, so long V_1 remains positive. Fig.1.27 elaborates this situation. Now line 3 is more inclined than line 1, and this represents that a negative charge on conductor B has increased the PD between the two conductors. In this situation, Eqns.1.34 assume the form

$$\left.\begin{aligned} Q_1 &= C_1V_1 + C(V_1 + V_2) = C_1V_1 + CV \\ -Q_2 &= -C_2V_2 - C(V_1 + V_2) = -C_2V_2 - CV \end{aligned}\right\} \qquad \text{................ (1.35)}$$

A very large value of $-Q_2$ will pull down V_1 to negative level.

In the above derivations, some given amounts of charges (as cause) were imparted to the two conductors, and their potentials (as effect) were derived. As such, these derivations are classed under what can be termed as '**Q- concept**'. Same results (Eqns.1.34 to 1.35) can also be obtained with '**V- concept**', wherein the charges (as effect) on the two conductors are derived in terms of their potentials (as cause). Figs.1.28(a) and (b) illustrate the topic with this V- concept, respectively for similar and opposite charges on the two conductors, and they correspond to Figs.1.26(d) and 1.27(a). Thus, in Fig.1.28(a), assuming $V_1 > V_2$, we have

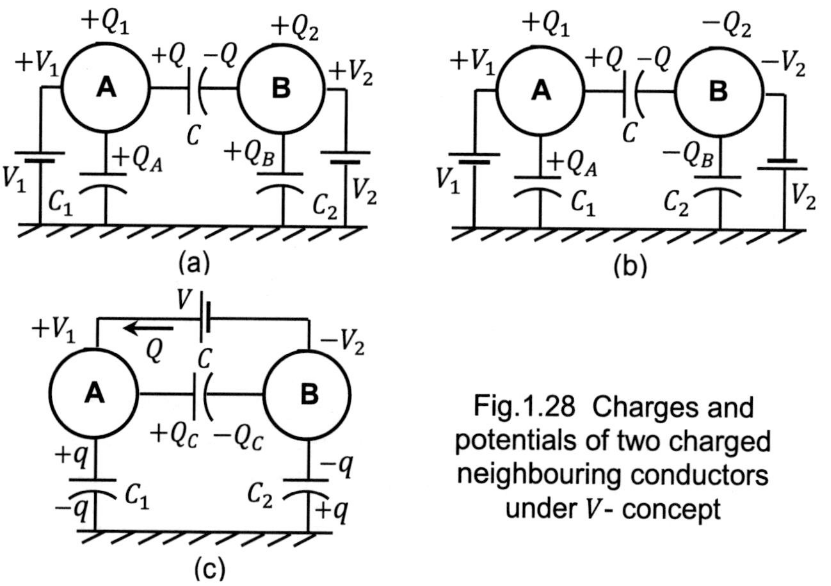

Fig.1.28 Charges and potentials of two charged neighbouring conductors under V- concept

$$\left.\begin{aligned} Q_1 &= Q_A + Q = C_1V_1 + C(V_1 - V_2) \\ Q_2 &= Q_B - Q = C_2V_2 - C(V_1 - V_2), \end{aligned}\right\} \quad \text{which are Eqns.1.34.}$$

And in Fig.1.28(b),

$$\left.\begin{aligned} Q_1 &= Q_A + Q = C_1V_1 + C\{V_1 - (-V_2)\} = C_1V_1 + C(V_1 + V_2) \\ -Q_2 &= -Q_B - Q = C_2(-V_2) - C(V_1 + V_2), \end{aligned}\right\} \quad \text{which are Eqns.1.35.}$$

The treatment with the V- concept is thus much simpler than that with the Q- concept.

If a voltage V is applied to a capacitor, Fig.1.28(c), considering the earth capacitances of its conductors, the two parallel circuits across the source are C_1 and C_2 in series, and C. As the conductors are oppositely charged, this case is similar to that of Fig.1.28(b), and the charge distribution is shown accordingly. If V_1 and $(-V_2)$ be the potentials of A and B respectively, with respect to earth, the magnitudes of V_1 and V_2 are derived as below.

$$q = C_1V_1 = C_2V_2 \text{ and } V = V_1 + V_2. \text{ Hence, } \frac{V_1}{V_2} = \frac{C_2}{C_1} \text{ or, } \frac{V_1}{V_1+V_2} = \frac{C_2}{C_1+C_2}$$

$$\text{or, } V_1 = V\frac{C_2}{C_1+C_2} \text{ and similarly, } V_2 = V\frac{C_1}{C_1+C_2}$$

Division of V into V_1 and V_2 is thus decided by the earth capacitances C_1 and C_2, and the mutual capacitance C does not interfere. V_1 and V_2, therefore, cannot be thought of without the earth capacitances C_1 and C_2. This theoretical situation is ideal, and does not prevail in practice owing to the finite insulation resistances of the conductors to earth, and the topic will be supplemented in Sec.2.10.

Total charge drawn from the source

$$Q = Q_C + q = CV + V\left[\frac{1}{\frac{1}{C_1}+\frac{1}{C_2}}\right]$$

As C_1 and C_2 decrease, the term within the square bracket approaches zero, and as such, $Q \approx CV$. In practice, earth capacitances are usually negligible with capacitance values (C) of the order of $0.1\mu\text{F}$ and above. Eliminating V_1 from Eqns.1.34, V_2 can be expressed in the form $\boldsymbol{V_2 = K_1Q_2 + K_2}$, the K's being constants. The V_2-Q_2 relationship is thus linear, as shown at Fig.1.29(a). The V-Q_2 relationship, Fig.1.29(b), is also linear but comprises of two parts for $V_1 \geq V_2$ and $V_2 \geq V_1$. We consider them separately.

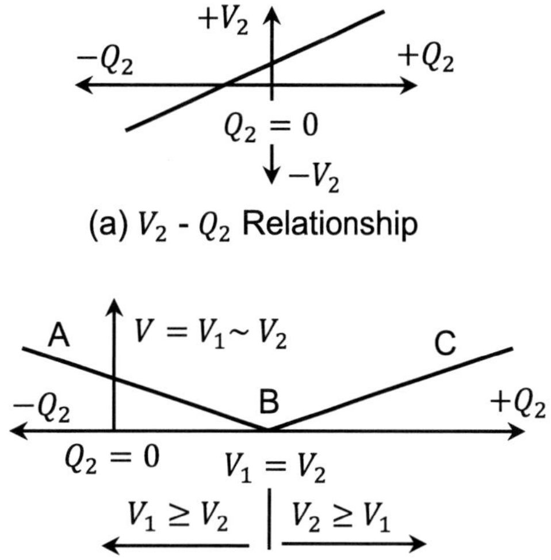

Fig.1.29 Potentials and PD of two neighbouring charged conductors

a) $\boldsymbol{V_1 \geq V_2}$**:** As was discussed above, this happens as Q_2 grows from negative to positive. $V = V_1 - V_2$, and Eqns.1.34 apply.

$$\left.\begin{aligned} Q_1 &= C_1V_1 + CV \\ Q_2 &= C_2V_2 - CV \end{aligned}\right\}$$

$$\text{or, } \left.\begin{aligned} (C_2/C_1)Q_1 &= C_2V_1 + (CC_2/C_1)V \\ Q_2 &= C_2V_2 - CV \end{aligned}\right\}$$

$$\begin{aligned} \text{Hence, } (C_2/C_1)Q_1 - Q_2 &= C_2(V_1 - V_2) + (CC_2/C_1)V + CV \\ &= C_2V + (CC_2/C_1)V + CV \\ &= [C_2 + (CC_2/C_1) + C]V \end{aligned}$$

or, $K_1 - Q_2 = K_2V$ (all K's are constants)

or, $\boldsymbol{V = -K_3Q_2 + K_4}$. This gives the line AB.

b) $\boldsymbol{V_2 \geq V_1}$**:** This happens with positive values of Q_1 and Q_2. $V = V_2 - V_1$, and Eqns.1.34(a) apply.

$$\left.\begin{aligned} Q_1 &= C_1V_1 - CV \\ Q_2 &= C_2V_2 + CV \end{aligned}\right\}$$

$$\text{or, } \left.\begin{aligned} (C_2/C_1)Q_1 &= C_2V_1 - (CC_2/C_1)V \\ Q_2 &= C_2V_2 + CV \end{aligned}\right\}$$

$$\begin{aligned} \text{Hence, } Q_2 - (C_2/C_1)Q_1 &= C_2(V_2 - V_1) + (CC_2/C_1)V + CV \\ &= C_2V + (CC_2/C_1)V + CV \\ &= [C_2 + (CC_2/C_1) + C]V \end{aligned}$$

or, $Q_2 - K_1 = K_2V$

or, $\boldsymbol{V = K_3Q_2 - K_4}$ and this gives the line BC.

1.11 Polarization of Dielectric: Electrostatic induction and electric polarization are similar phenomena, but they occur differently in matter; the former occurs in conductors whereas the latter in dielectrics. However, the term polarization is often used to mean induction in conductors. The subject is better explained by the microscopic (atomic) composition of matter. A very simple and crude conception of the same follows. The radii of atoms are of the order of the atomic unit of length, the angstrom (Å = 10^{-10} m), whereas those of electrons and nuclei are of the order of only 10^{-5} Å (10^{-15} m). With classical mechanical view, therefore, an atom is essentially 'empty'. With more appropriate wave mechanical view, however, the fast-moving electrons behave as continuous charge distribution, creating thereby a cloud-like atmosphere, surrounding the nucleus. Such representation of electrons is referred to as 'electron cloud'. A mass of matter is thus considered as composed of innumerable number of nuclei embedded within the electron cloud. Conductors and dielectrics, as we know, differ in the type of the chemical bond that holds their (like or unlike) atoms together.

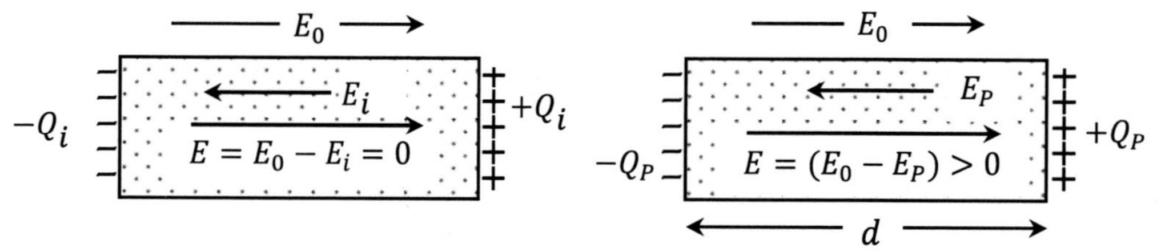

(a) Electrostatic induction (b) Electric polarization

Fig.1.30 Electrostatic induction and electric polarization

In (metallic) conductors, the valence electrons are 'free' in the sense they do not remain confined to their parent atoms; they belong to, or are shared by, all the atoms in a mass. A conductor is thus considered as composed of heavy positive ions consisting of the nuclei and the 'bound cloud' (electron cloud associated with the 'bound' electrons), lying within the light 'valence cloud' (electron cloud associated with the 'valence' electrons). Under no external electric field, Coulomb forces hold the valence cloud and the positive ions together (metallic bond). An external electric field E_0, may be due to proximity of another charged conductor, shifts the negatively charged valence cloud in the conductor along a direction opposite to that of the external field. As such, the two opposite ends of the conductor in Fig.1.30(a) acquire some equal and opposite charges $\pm Q_i$, and we call this 'electrostatic induction'. Even a small external field is sufficient to induce appreciable amount of charges in conductors. Quantitatively, these charges create an opposing field $E_i = Q_i/(\varepsilon_0 A)$ within the conductor, A being the sectional area of the conductor. Under equilibrium, E_i is just equal and opposite to E_0, so that the net field inside the conductor ($E = E_0 - E_i = 0$) is zero, which is obvious for a conductor. As E_0 increases, $Q_i(= \varepsilon_0 A E_0)$ increases linearly with it, and this implies shifting of more valence electrons. Now, the question arises what happens after all the valence electrons have been shifted. The same question also arises in cases like the charging of capacitor by DC source, which also involves shifting of valence electrons from positive to negative plate. The single answer to all such questions is that the number of valence electrons, even in a very small mass of conductor, is so huge that it cannot be exhausted, at least by practical values of the field strength.

In the case of dielectrics, the valence electrons are not free; they are also bound like other electrons. Hence, the entire electron cloud is bound cloud only. Chemical bonding generally occurs either between unlike atoms in the form of positive and negative ions held together by Coulomb forces (Ionic bonds, as in NaCl), or between like atoms through sharing of their valence electrons (Valence bonds, as in diamond, silicon, germanium and hydrogen). In both the cases, the valence electrons remain confined to the nearest neighbour atoms only, and their mobility is thus restricted. Under the influence of some external electric field E_0, the centre of symmetry of the electron cloud shifts from its original position on the centre of symmetry of the nuclei in a mass (in the tune of only 10^{-17} m, which is very small compared to the radius of the atom). Hence, same as with conductors, some equal and opposite charges $\pm Q_P$ appear on the opposite ends of the dielectric, which is called 'electric polarization', Fig.1.30(b). Unlike in conductors, a much larger external field is required to produce any appreciable amount of polarized charge in a dielectric. The polarized charges create an opposing field $E_P = Q_P/(\varepsilon_0 A)$ in the dielectric, but in this case the net field in the dielectric is positive $(E = (E_0 - E_P) > 0)$. The shifting of the bound cloud, as in dielectrics, also occurs in conductors, but the effect is insignificant compared to that of the valence cloud. The essential difference between induction and polarization is, therefore, that the former produces appreciable amount of charge in conductors even by some feeble external field, whereas the latter produces small amount of charge in dielectrics even with strong external field. There are other distinctions also. Induced charges may leave the mass of the conductor, but polarized charges cannot leave the mass of the dielectric. Hence, a conductor can be charged by induction, but a dielectric cannot be charged by polarization.

The electron cloud shifting mechanism of polarization discussed above is called 'electronic' polarization; this is the only mechanism that occurs in monoatomic inert gases. In polyatomic molecular gases, two other mechanisms, called 'orientational' polarization and 'ionic' polarization, also occur in addition. Moreover, the internal fields in liquid and solid dielectrics complicate the subject further. The total polarization is contributed by all these components. These details, however, are not necessary at this stage of study.

We now discuss the consequence of polarization. In Fig.1.30(b), let d be the length of the dielectric along the direction of the external electric field. The surface density of the polarized charge per unit cross sectional area of the dielectric is defined as the quantity '*electric polarization*', denoted by P. Defined as such, polarization is induced charge density, and so is a scalar quantity. Hence,

$$P = \frac{Q_P}{A} = \frac{Q_P.d}{A.d} = \frac{p_e}{V} \quad \text{.............................. (1.36)}$$

where $p_e = Q_p.d$ is the magnitude of the electric dipole moment vector of the polarized volume $V = A.d$ of the dielectric. Therefore, the polarization is also defined as the induced dipole moment per unit volume of a dielectric. With the usual assumption that the dielectric material is isotropic (materials exhibiting the same physical and other properties measured in all directions are called isotropic materials), the entire volume of the dielectric is uniformly polarized, that is, the polarization P at all points in the volume of the dielectric remains the same in a uniform or parallel field. If the field is not uniform, e.g., a radial field, the polarization is not uniform, even if the dielectric is isotropic. It then varies from point to point in the volume of the dielectric, and at any point it is given by dp_e/dV, where dp_e is the dipole moment of an elementary volume $dV = dl.dA$ at that point. Therefore, electric field is the cause and polarization is its effect. Except for some few 'ferroelectric' materials (which exhibit appreciable 'hysteresis' property similar to that observed in 'ferromagnetic' materials, Sec.6.1), electric polarization is a reversible process, that is, the induced charges disappear as soon as the external field is withdrawn; this equally applies to electrostatic induction in conductors.

It will now be explained what exactly happens when a dielectric is subjected to an electric field. Two alternative concepts can be developed under two following situations.

(A) Q – concept:

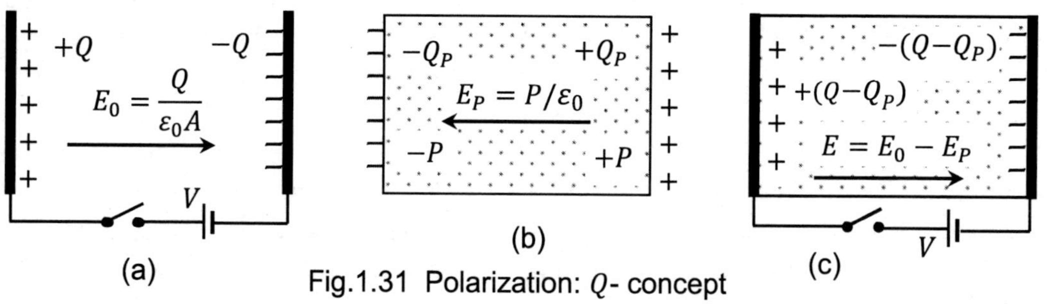

Fig.1.31 Polarization: Q- concept

In Fig.1.31(a), the parallel plate capacitor of area A and inter-plate distance d is first charged to the battery voltage V and to charges $\pm Q$ with vacuum (or air) as the dielectric. The battery is then switched off, and the capacitor retains the charges $\pm Q$. With the usual assumption that $A \gg d$, the uniform charge density on the plates creates a parallel field between the plates. The electric field strength at any point between the plates, as was derived in Sec.1.7.2, $E_0 = Q/(\varepsilon_0 A)$. Let us then fill the space between the plates with some dielectric material, as in Fig.1.31(b). Under the influence of the electric field, the dielectric is polarized and the polarities of the polarized charges $\pm Q_P$ being opposite to those of the main charges $\pm Q$ on the plates, the net charges on the plates reduce to $\pm(Q - Q_P)$, Fig.1.31(c). The polarization field $E_P = \frac{Q_P}{\varepsilon_0 A} = \frac{P}{\varepsilon_0}$ acts in opposition to the main field E_0, and thus reduces the latter to $E = E_0 - E_P$. This explains how the electric field strength reduces from its free-space value E_0 to E in any dielectric. However, the amount of flux from the positive plate to the negative plate $\Psi\,(= Q)$ and the flux density at any point between the plates $D\,(= \Psi/A = Q/A)$ do not change due to the insertion of the dielectric in the field. This implies that the polarized charge is not associated with any flux. This may appear misleading, but is quite correct, as it follows from the arbitrary definition of the flux that the amount of flux from a given amount of charge remains the same in all media.

Again, we know that the flux density at any point in the field is given by $D = \varepsilon_0 E_0 = \varepsilon E$. Hence, we get $D = \varepsilon_0 E_0 = \varepsilon_0(E + E_P) = \varepsilon_0 E + \varepsilon_0 E_P$

or, $D = \varepsilon_0 E + P$ .... (1.37)

The quantity P in Eqn.1.37 is, therefore, a flux density vector, whereas it is a charge density scalar in Eqn.1.36. Although the symbol is the same, the two quantities are different. Their numerical values are, of course, the same. In fact, the definition of polarization as the flux density vector is more common in textbooks than as the charge density scalar.

(B) E – concept:

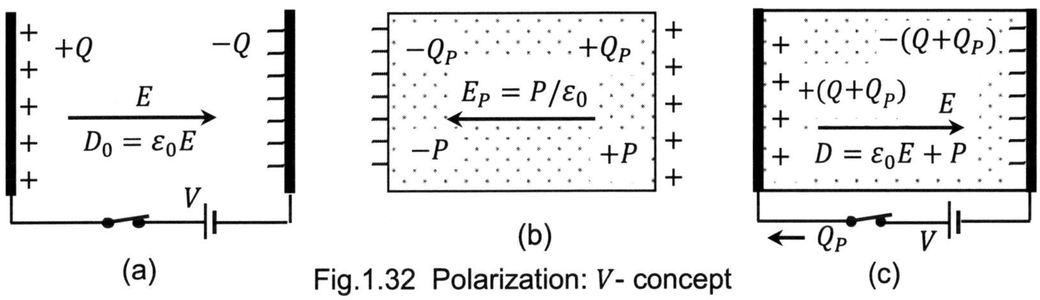

Fig.1.32 Polarization: V- concept

In this concept, Fig.1.32, the battery remains connected to the capacitor, and so the field strength between the capacitor plates is maintained at $E = V/d$. With vacuum as the dielectric, Fig.1.32(a), the situation remains the same as in the 'Q - Constant' concept. The flux density between the plates, $D_0 = Q/A = \varepsilon_0 E$. If then some dielectric material be inserted in the space between the plates, Fig.1.32(b), the dielectric will be polarized. The polarities of the induced charges $\pm Q_P$ being opposite to those of the main charges $\pm Q$, the field strength tends to reduce by $E_P = \frac{Q_P}{\varepsilon_0 A} = \frac{P}{\varepsilon_0}$. To maintain the field strength at $E = V/d$, an additional charge Q_P is drawn from the battery, that neutralises the polarized charges, Fig.1.32(c). As such, the total charges on the plates become $\pm(Q + Q_P)$ and the flux density in the inter-plate space increases from D_0 to

$$D = \frac{Q+Q_P}{A} = \frac{Q}{A} + \frac{Q_P}{A} = \varepsilon_0 E + P, \text{ which is Eqn.1.37}$$

Although parallel field configuration has been considered for derivation of Eqn.1.37, radial field configuration could also be considered, ending, of course, with the same result. Eqn.1.37 implies that the total flux density D in a dielectric may be thought of as comprising of two components – the flux density $\varepsilon_0 E$ due to the net field E, plus the polarized flux density P. Both the concepts lead to the same result and it remains the situation that selects adaptability of any particular concept. In Electrostatics, we found both the Q- concept and E- concept applicable. In Magnetism with Isolated Pole Theory, only the m- concept is applicable, which is analogous to the Q- concept, whereas in Electromagnetism, only the H- concept is applicable, which is analogous to the E-concept. These remarks will be appreciated in Sec.3.11.

1.12 Electric Susceptibility: At any point in a volume of dielectric (except for few ferroelectric dielectrics), the polarization charge density P, hence the polarization field strength E_P $(= P/\varepsilon_0)$ and the polarization flux density P, all these quantities are directly proportional to the electric field strength E at that point. The constant of proportionality defines a property of the dielectric material called its *electric susceptibility* χ_e. Consequently, the definition of electric susceptibility depends on the quantity chosen. The quantity *polarization field strength* E_P is usually chosen, and as such, the electric susceptibility is defined as the unit-less ratio

$$\chi_e = \frac{Polarization\ field\ strength,\ E_P}{Net\ field\ strength\ producing\ the\ polarization,\ E}$$

$$= \frac{P}{\varepsilon_0\, E} \qquad \text{......................................} (1.38)$$

We are now in a position to explain why the relative permittivity ε_r of any material is greater than its free-space value of unity. A dielectric differs from the vacuum in that whereas the former can be polarized, the latter cannot be, simply because there is no atom available in the vacuum for polarization. This fact logically prompts us to expect that there must be some relation between permittivity and susceptibility. The desired relationship is derived below.

Substituting for P from Eqn.1.38 in Eqn.1.37 we get

$$D = \varepsilon_0 E + P = \varepsilon_0 E + \varepsilon_0 E . \chi_e$$

$$= \varepsilon_0 E(1 + \chi_e)$$

We also have $D = \varepsilon E = \varepsilon_0 \varepsilon_r E$. Comparing these two relations, we get

$$\varepsilon_r = \chi_e + 1, \text{ or } \chi_e = \varepsilon_r - 1 \qquad \text{.............................} (1.39)$$

An alternative derivation of Eqn.1.39 follows. In Sec.1.3 we had $E_P = (\varepsilon_r - 1)E$, and we also have $E_P = P/\varepsilon_0$. Eliminating E_P from these two equations, we get $P = \varepsilon_0\,(\varepsilon_r - 1)E$. Again, Eqn.1.38 gives $P = \varepsilon_0\,\chi_e E$. Eliminating P from last two equations,

$$\chi_e = \varepsilon_r - 1, \text{ which is Eqn.1.39.}$$

For all substances, χ_e is some (unit-less) positive number, and hence ε_r is always greater than one. This explains why the relative permittivity ε_r of any material is greater than its free-space value of unity. These remarks are equally valid for all solid, liquid and gaseous dielectrics.

Sometimes the electric susceptibility is defined as the ratio

$$\chi_e = \frac{Polarization\ flux\ density,\ P}{Net\ field\ strength\ producing\ the\ polarization,\ E} = \frac{P}{E}$$

When so defined, electric susceptibility relates to permittivity and relative permittivity by the relation $\chi_e = \varepsilon - \varepsilon_0 = \varepsilon_0\,(\varepsilon_r - 1)$ and it bears the same unit of permittivity, F/m.

Polarizability is a related quantity commonly used in the atomic study of dielectrics. It is a property of atoms and molecules. In macroscopic terms, it may be expressed as

$$\frac{Polarization\ dipole\ moment,\ Q_P.d}{Net\ field\ strength\ producing\ the\ polarization,\ E}$$

$$= \frac{Q_P\, Ad}{AE} = \frac{P}{E}(volume) = \varepsilon_0.\chi_e.(volume) \text{ (by Eqn.1.38)}$$

As this refers to some volume of dielectric, macroscopic polarizability is not a material property. However, polarizability per unit volume $(= \varepsilon_0.\chi_e)$ is a material property. Polarizability is expressed in the unit of farad-metre^2 (F.m^2). In electrical power engineering, the properties susceptibility and polarizability are seldom used; rather the permittivity is one prime property.

1.13 Force Between Electrified Conductors: Coulomb's law gives the force between very small (point-like) charged bodies. Viewed with a broader sense, Coulomb's law incorporates the implied condition that the distance between the bodies is very great compared with their dimensions. The law, therefore, does not directly apply for neighbouring charged bodies, even when they are conductors and spherical in shape. When charged bodies of appreciable sizes lie in proximity, they polarize each other by induction, and Coulombian forces act between these polarized charges. The amount of polarized charges in dielectrics are very small (as was discussed in Sec.1.11), and the force between electrified dielectrics remains too insignificant and impracticable to address. In the case of conductors, however, the induced charges being quite appreciable, the force between neighbouring charged conductors needs consideration. Numerous scientists and researchers, starting from Poisson, then Lord Kelvin (Sir William Thomson, in 1845), and Maxwell (in 1891), investigated the subject of the force between two neighbouring charged conductors (of the simplest shape of sphere). They analysed the subject with different methodologies, and derived various formulae under different conditions, using higher mathematics. Although the cumbersome quantitative details of these investigations need not be considered at this stage of study, some simple and crude qualitative concept of the subject is essential, and this can be developed from our previous discussion in Sec.1.10.5 on the charges and potentials of two neighbouring conductors. Thus in Fig.1.33(a), the charges Q_1 and Q_2 on the two neighbouring conductors A and B, split into the following parts

$$Q_1 = Q_A + (+Q) \text{ and}$$
$$Q_2 = Q_B + (-Q)$$

decided by the earth and mutual capacitances C_1, C_2 and C of the conductors. Four charges can be seen in the figure on the two conductors, and their possible combinations generate four forces between the conductors. These forces can be designated as

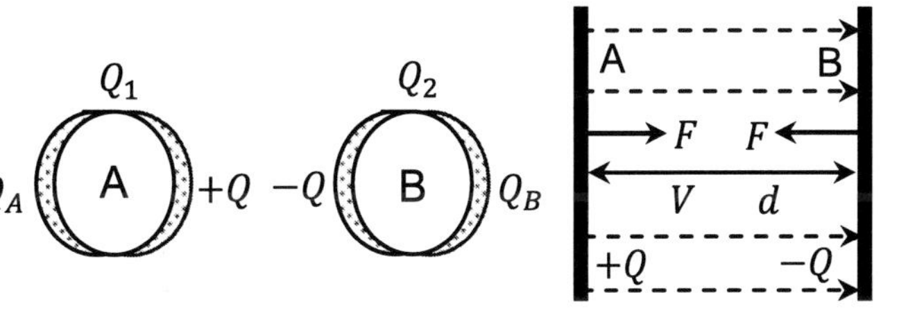

(a) Charge distribution in electrified conductors (b) Attractive force in parallel plate capacitor

Fig.1.33 Force between electrified conductors

F: Between $(+Q)$ and $(-Q)$
F_1: Between Q_A and $(-Q)$
F_2: Between Q_A and Q_B
F_3: Between $(+Q)$ and Q_B

These forces may be attractive, repulsive, or even zero, depending on the polarity and magnitude of the charges, and their resultant gives the force between the two conductors. What mainly complicates the subject is that these charges are not concentrated ones but distributed over the conductor surfaces with varying densities as very roughly indicated by the width of the dotted areas in the figure. Let us first consider the attractive force F alone (ignoring the other forces). In the simplest case, the attractive force between the plates of a charged parallel plate capacitor, Fig.1.33(b), was derived in Sec.1.8.5 as $F = Q^2/(2\varepsilon A)$. Since the capacitance of the capacitor $C = Q/V = \varepsilon A/d$, where V and d are respectively and the PD and the distance between the plates, F can be expressed in various forms

$$F = \frac{Q^2}{2\varepsilon A} = \frac{QV}{2d} = \frac{CV^2}{2d} = \frac{Q^2}{2Cd} \qquad \text{.......................... (1.40)}$$

[Imagine now what may be the force if the dielectric happens to be composite one. For the parallel plate capacitor with composite dielectric of Fig.1.23(b) to be equivalent to a single-dielectric capacitor with the permittivity of the dielectric ε, having the same capacitance C and the same overall dimensions, it follows from Eqn.1.32

$$C = \frac{A}{\frac{d_1}{\varepsilon_1} + \frac{d_2}{\varepsilon_2} + \frac{d_3}{\varepsilon_3} + \cdots \frac{d_n}{\varepsilon_n}} = \frac{A}{\frac{d}{\varepsilon}}$$

or, $$\frac{d}{\varepsilon} = \frac{d_1}{\varepsilon_1} + \frac{d_2}{\varepsilon_2} + \frac{d_3}{\varepsilon_3} + \cdots \frac{d_n}{\varepsilon_n} \qquad \text{..................... (1.41)}$$

where, $d = d_1 + d_2 + d_3 + \cdots d_n$ is the thickness of the composite dielectric. As such, Eqn.1.40 remains valid for the present case of composite dielectric also, but with the values of d/ε and C substituted from Eqn.1.41.]

We now consider the attractive force F between the two conductors A and B in Fig.1.33(a). What can be generalised from Eqn.1.40, the proportionality $F \propto V^2$ applies to the capacitance forming charges $\pm Q$ on the two neighbouring conductors of given sizes and shapes so long they do not change their physical positions. As was seen in Fig.1.29(b), Sec.1.10.5, the V- Q_2 relationship is of the form $V = AQ_2 + B$, where the constants A and B have opposite signs. Hence, F-Q_2 relationship is a quadratic equation of the form $F = aQ_2{}^2 - bQ_2 + c$, where a, b, and c are positive constants. The F-Q_2 curve is, therefore, the parabola shown at Fig.1.34(a). Repulsive force between like charges being positive, it is customary to represent repulsion along positive direction of rectangular axes. The inverted curve F in Fig.1.34(b) only complies to this practice.

We now come to the rest three forces F_1, F_2 and F_3; let F_{13} be their resultant. These forces also vary with Q_2. Although the force between distributed charges cannot be computed by simple mathematics, some idea of the same can, however, be developed logically from our discussion in Sec.1.10.5, if we assume the charges as concentrated and apply Coulomb's law.

(a) Firstly, with conductor B uncharged ($Q_2 = 0$), $Q_B = +Q$, and as such, attraction F_1 and repulsion F_2 almost neutralise, so that the resultant F_{13}($\approx$ repulsion F_3) is positive. The charges $\pm Q$ being more nearer, however, F is more negative, so the net F_R is negative.

(b) Secondly, with increasing values of $+Q_2$, both $+Q_A$ and $+Q_B$ increase and $\pm Q$ decreases. As such, attraction F_1 decreases, repulsion F_2 increases, and repulsion F_3 does not change to comparable extent. Net result is that repulsion F_{13} increases with increasing values of $+Q_2$.

(c) As third clue, for large values of $-Q_2$, Q_B is negative; this makes all the forces F_1, F_2 and F_3 individually attractive, so F_{13} is negative, and this attraction increases with increase in $-Q_2$. This means, F_{13} must be zero for some negative value of Q_2, below which it is attractive and above which repulsive.

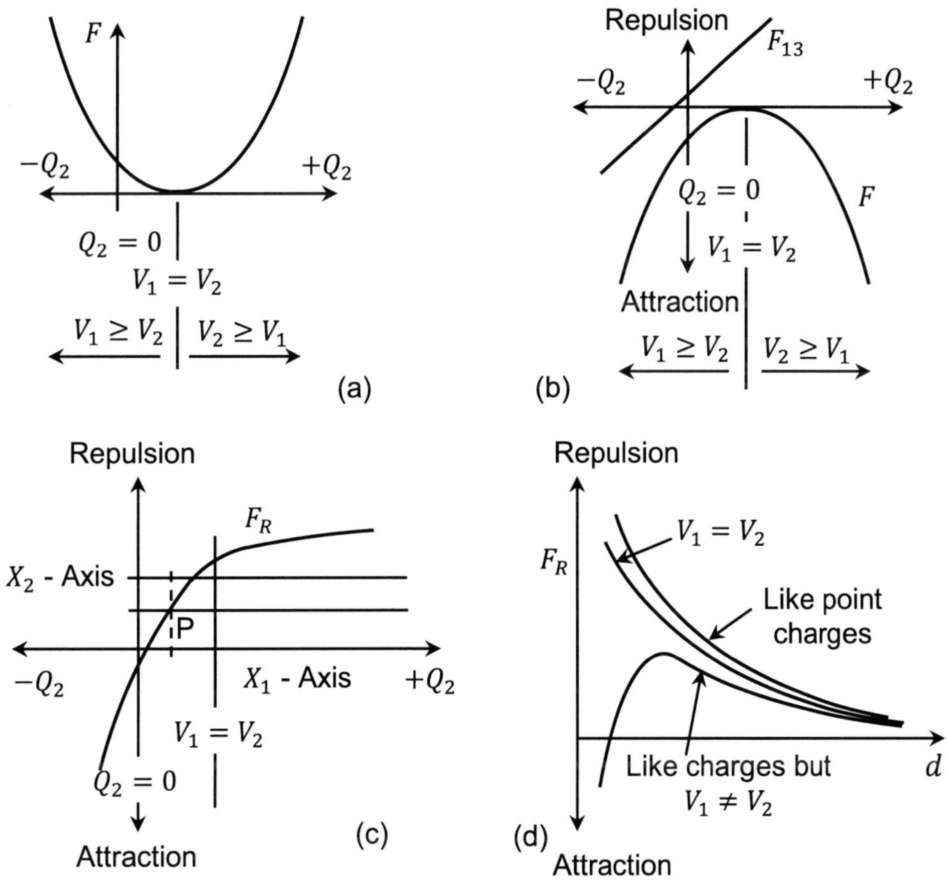

Fig.1.34 Force between charged conductors

Based on these observations, and approximating for our present purpose that F_{13}-Q_2 relation is linear, it can be represented by the straight line F_{13} in Fig.1.34(b). The resultant force F_R between the conductors, resultant of F and F_{13}, can thus be predicted to be of the general shape shown in Fig.1.34(c). Exact profile of the curve, and its exact placement in the frame of the rectangular coordinate axes, will depend, of course, on all the three capacitances, and so on all the factors those decide these capacitance values, mainly the distance of the conductors, with which the forces vary with square law. Thus, for some given value of $+Q_2$ shown by the dotted vertical line in Fig.1.34(c), F_R may be positive with X_1- axis, negative with X_2- axis, and zero with the X- axis passing through the point P. It can be seen that some amount of $+Q_2$ is necessary in any case to make F_R zero.

So far, we assumed that the distance between the conductors is constant. As the charges ($\pm Q$) are the nearest ones amongst all the pairs behind the four forces, F is most sensitive to the change in the distance between the conductors than the rest three forces are. Variation of the forces with the distance d between the conductors, with given set of values of Q_1, Q_2, C_1 and C_2, can be observed under the following situations.

A. Conductors at same potential, $V_1 = V_2$: This, of course, happens with like charges on the conductors. Here, $Q_A(= Q_1)$ and $Q_B(= Q_2)$ are like charges and Q is zero. So, F_R is repulsive for all values of d, Fig.1.34(d). As the conductors come closer, Q_A and Q_B are pushed apart further, so the F_R curve for $V_1 = V_2$ lies below the curve for the point charges.

B. Conductors at different potentials, $V_1 \neq V_2$: This is the general situation, of which situation A is only a special case. Before we discuss the forces, let us first analyse the capacitance, charge and potential variations associated with this general situation.

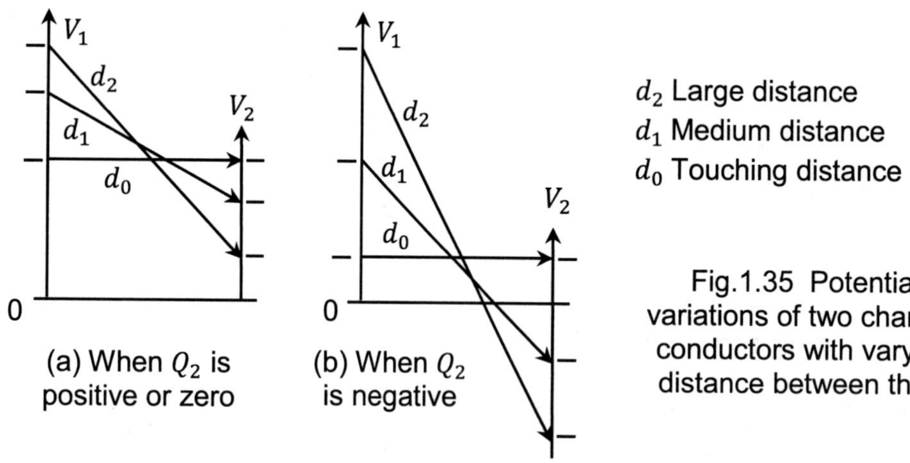

(a) When Q_2 is positive or zero

(b) When Q_2 is negative

Fig.1.35 Potential variations of two charged conductors with varying distance between them

We have in general

$$Q = CV = C(V_1 - V_2) = C\left[\frac{Q_A}{C_1} - \frac{Q_B}{C_2}\right] = C\left[\frac{Q_1}{C_1} - \frac{Q}{C_1}\right] - C\left[\frac{Q_2}{C_2} + \frac{Q}{C_2}\right]$$

$$\text{or, } Q + \frac{CQ}{C_1} + \frac{CQ}{C_2} = \frac{CQ_1}{C_1} - \frac{CQ_2}{C_2} \quad \text{or, } Q\left[1 + \frac{C}{C_1} + \frac{C}{C_2}\right] = \frac{CC_2Q_1 - CC_1Q_2}{C_1C_2}$$

This simplifies to give an important identity

$$Q = \frac{C_2Q_1 - C_1Q_2}{(C_1C_2/C) + C_1 + C_2}$$

As one limiting case, when the distance between the conductors is infinitely large, $(d = \infty)$, $C = 0$, $Q = 0$. This simply explains that no induction occurs between distant lying conductors. In this limiting case, $V_1 = \frac{Q_1}{C_1}$, $V_2 = \frac{Q_2}{C_2}$ and $V = V_1 - V_2 = \frac{Q_1}{C_1} = \frac{Q_2}{C_2}$. This is the maximum value of V. In another limiting case, as the distance between the conductors approaches zero (conductors are just about to touch), $d \to 0$, $C \to \infty$ and $Q \to \frac{C_2Q_1 - C_1Q_2}{C_1 + C_2}$. This is the maximum possible value of Q. If $C_1 = C_2$, as for two equal spheres at equal height above the earth surface, $Q \to \frac{Q_1 - Q_2}{2}$. At this touching distance, conductors *approach* to become equipotential, $V \to 0$, $C \to \infty$ and $Q = CV$ complies. Same condition prevails with the capacitance C_E in Fig.1.25(a). Irrespective of Q_2 being positive, negative or zero (with Q_1 positive), as the conductors come closer, Q grows to its above touching value; V_2 increases, V_1 decreases, so $V \to 0$, as explained by Fig.1.35.

Upon eventual touching of the conductors, true equipotential condition discussed in situation A above is *achieved* without any effective charge transference between the conductors; only the charges $\pm Q$ neutralise, the potential of the united conductor becomes $(Q_1 + Q_2)/(C_1 + C_2)$ which may be positive, negative or zero, and the forces are as explained in situation A above.

Based on the above discussion, the forces between the conductors, *before the equipotential state is achieved*, can be observed as below:

(a) One conductor charged and another uncharged ($Q_2 = 0$): Attraction occurs at all distances. As d decreases, repulsion F_{13} increases, and attraction F increases faster than repulsion F_{13}. Hence, the net attraction F_R increases.

(b) Conductors are unlikely charged: Attraction occurs at all distances and increases with decreasing distance.

(c) Conductors are likely charged: This is an interesting situation. As the conductors come closer, attraction F gradually dominates over the repulsion F_{13}. The net force F_R, as shown in Fig.1.34(d), rises upto some maximum repulsion at some small distance, then drops sharply from repulsion to attraction at very close distance. This means, depending on the distance, F_R may be repulsive, attractive, and so zero also, which is in agreement with what we discussed with Fig.1.34(c).

(d) Force between a charged conductor and the earth: It reveals from the concept of electrical image discussed in Sec.1.10.5, Fig.1.25(b) that the earth exerts an attractive force F_R on a charged conductor, and the force increases with decreasing height of the conductor above earth. Earth capacitances being very small, such forces are very feeble.

(e) One conductor charged and another earthed: In Fig.1.33(a), if one conductor, say B, be earthed, Q_B becomes zero. There are then two forces between the two conductors, both attractive. Hence, F_R is attractive, and this increases with decreasing distance.

Above discussion developes the concept that with decreasing distance between two charged conductors, F_{13} line in Fig.1.34 is raised, F parabola narrows and F_R curve lowers.

Before concluding this section, one significant point requires mention. The forces F_1, F_2 and F_3 between the two conductors arise of the charges Q_A and Q_B, and these charges appear due to the earth capacitances C_1 and C_2 of the two conductors. As earth capacitances of practical conductors are generally too low compared to their mutual capacitance, the usual assumption that generally remains implied, F is the only force between the conductors.

1.14 Energy Stored in Electric Field: Whereas the electrostatic potential energy PE_E of a charged conductor, Sec.1.8, is due to its position in an electric field, the creation of any electric field itself involves expenditure of energy. *The amount of work done or energy expended by some external agency in creating an electric field is stored as electrostatic potential energy SE_E in the field.* Whereas PE_E is possessed by some charged conductor by virtue of its position in an electric field, SE_E is possessed by the electric field itself. They, therefore, differ by definition and conception, although both represent potential energy, and so are measured in terms of electric potential or PD. Energy stored in electric field SE_E can be derived with alternative concepts and considerations. Although we continue the discussion on SE_E with reference to a capacitor, it remains equally valid for conductor as well.

(A) Consideration – 1 (Q – concept): Charging of a capacitor may be thought of as gradual transference of infinite number of elemental charges from one plate of the capacitor to another *through the dielectric.* As opposite charges on the capacitor plates grow, the field between the plates becomes stronger, and so, more and more work is done by the external agency in transferring the charge against the increasing field strength. If charges $\pm Q$ on the plates raise the PD between the plates to V, we know $V = Q/C$, where C, the capacitance of the capacitor, is constant. So, V increases linearly with Q. The V-Q curve is an inclined straight line as shown at Fig.1.36(a). At any instant of time t during the process of charging the capacitor, if $\pm q$ are the instantaneous cumulative charges on the plates, which raise the instantaneous PD to v, we have $v = q/C$. With this potential difference v, if dq amount of charge is transferred in next elemental time dt, the work done or increase in SE_E during time dt is given by

$$d(SE_E) = v.dq \qquad \text{.................................} \quad (1.42)$$

represented by the area of the elemental strip shown, potential or PD being work done per unit charge transferred. The SE_E in the electric field of the capacitor with charges $\pm Q$ and potential difference V is, therefore,

$$SE_E = \int_{q=0}^{q=Q} d(SE_E) = \int_{q=0}^{q=Q} v.dq = \frac{1}{C}\int_{q=0}^{q=Q} q.dq = \frac{1}{C}\left[\frac{q^2}{2}\right]_0^Q = \frac{VQ}{2}$$

which is the area of the dotted triangle OAQ *below* the V-Q curve. In the process of charging a capacitor, or a conductor, work is always done by some external agency *against the field* in moving the charge, and such work done is stored as SE_E in the field.

During discharging, the field releases the energy that was stored previously. Hence, SE_E is always positive, but cannot be negative, whereas PE_E can be positive or negative, as was explained earlier (Sec.1.8). This is one feature the SE_E differs from the PE_E. The principal distinction between PE and SE, be they electric or magnetic, is that whereas the former describes the ability to deliver mechanical energy, the latter describes the ability to deliver electrical energy.

It must be noticed that *the SE_E in the attractive field between two conductors in mutual capacitance* $(+VQ/2)$ *is the same as the PE_E of either conductor in the field of, and with zero reference at, the other conductor,* as was derived in Sec.1.8.5. Hence, if the distance x between the two conductors varies, we have

$$\frac{dSE_E}{dx} = \frac{dPE_E}{dx} \; (= -F)$$

It will be seen in Sec.5.7 (Eqn.5.24) that electromagnetic fields behave differently, so are not analogous to electric fields in this aspect.

(B) Consideration – 2 (Q – concept): During the process of charging the capacitor, as the PD across the capacitor v increases from zero to V uniformly (linearly) with the charge transferred, its average value is $V/2$. So, work done in transferring the charge Q under this average PD, which is the SE_E in the field, is given by $SE_E = VQ/2$. Using the relation $C = Q/V$, the SE_E with Q - concept can be expressed in the alternative forms

$$SE_E = \frac{VQ}{2} = \frac{CV^2}{2} = \frac{Q^2}{2C} \qquad \text{..................................} \; (1.43)$$

Eqn.1.43 also applies to the SE_E in the field surrounding an isolated charged conductor of any shape including the sphere, with charge Q, potential V and capacitance $C = Q/V$.

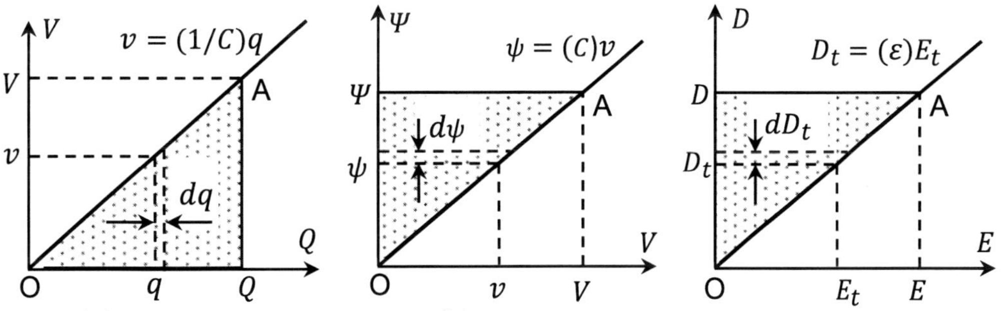

Fig.1.36 Energy stored in electric field

One important point needs elaboration at this stage. The factor ½ appears in Eqn.1.43 owing to the fact that V is a function of Q. During the charge transference process, the potential does not remain constant but increases linearly with the charge, which is the case in Electrostatics. In Current Electricity, however, the charge flow occurs under some fixed potential difference V, which is independent of the amount of charge flow. Therefore, in the latter situation, the work done is given by $W = \int dW = \int V.dq = V\int_{q=0}^{q=Q} dq = VQ$, and the factor ½ does not appear.

In the case of two neighbouring charged conductors with appreciable earth capacitances lying in proximity, Fig.1.28(c), the total SE_E of the system is the sum of the SE_Es in the three capacitances C, C_1 and C_2, and is given by

$$SE_E = \frac{1}{2}C_1{V_1}^2 + \frac{1}{2}C_2{V_2}^2 + \frac{1}{2}CV^2 \quad \text{.......} \quad (1.44)$$

This may be expanded as

$$SE_E = \frac{1}{2}C_1{V_1}^2 + \frac{1}{2}C_2{V_2}^2 + \frac{1}{2}C(V_1 \mp V_2)^2$$

(Negative sign applies when V_1 and V_2 are of similar sign, both positive or both negative, and positive sign applies when they are of opposite sign)

$$= \frac{1}{2}C_1{V_1}^2 + \frac{1}{2}C_2{V_2}^2 + \frac{1}{2}C({V_1}^2 + {V_2}^2 \mp 2V_1V_2)$$

$$= \frac{1}{2}(C_1 + C){V_1}^2 + \frac{1}{2}(C_2+C){V_2}^2 \mp CV_1V_2) \quad \text{.......} \quad (1.44(a))$$

The objective of writing the equation in this form will become evident in Sec.5.6.5.

(C) Consideration – 3 (Ψ – concept): Electric flux Ψ between the capacitor plates being numerically equal to the charges $\pm Q$ on the capacitor plates, Eqn.1.42 may be written as

$$d(SE_E) = v.d\psi \quad \text{.......} \quad (1.45)$$

with graphical representation as in Fig.1.36(b), wherein the instantaneous flux ψ $(= q = C.v)$ increases linearly with v. As such,

$$SE_E = \int dSE_E = \int v.d\psi = \frac{V\Psi}{2}$$

$$= \left[\frac{PD.flux}{2}\right] \quad \text{.......} \quad (1.46)$$

and is given by the area of the dotted triangle OAΨ *above* the Ψ- V curve.

These two alternative concepts sense differently. The former concept implies that the charge (cause, along X- axis) results in the electric PD (effect, along Y- axis). On the other hand, the latter concept implies that the electric PD (cause, along X- axis) results in the electric flux (effect, along Y- axis). The former concept implies that work has to be done, or energy has to be expended, in creating opposite charges on the capacitor plates. The latter concept implies that energy has to be expended in creating or developing an electric flux.

Expanding the relation $d(SE_E) = v.d\psi$ (Eqn.1.45) as

$$d(SE_E) = v.d\psi$$

$$= (E_t.d).(A.dD_t)$$

$$= (A.d).(E_t.dD_t)$$

where E_t and D_t are respectively the instantaneous electric field and electric flux density at any point between the capacitor plates at the instant t, and $(A.d)$ is the volume of the field between the capacitor plates, the increase in energy stored per unit volume in the field during time dt

$$d(SE_e) = E_t.dD_t \quad \text{.......} \quad (1.47)$$

as represented graphically in Fig.1.36(c). Here, D_t $(= \varepsilon.E_t)$ increases linearly with E_t. Hence, energy stored per unit volume of the field with field strength E and flux density D

$$SE_e = \int d(SE_e) \quad = \int E_t . dD_t = \frac{ED}{2}$$
$$= \left[\frac{field\ strength\ .flux\ density}{2}\right]$$

This is given by the area of the dotted triangle OAD *above* the D-E curve. This representation of energy stored per unit volume in terms of the area of the D-E curve helps deriving the hysteresis loss in ferroelectric materials in terms of the area of the hysteresis loop.

Using the relation $D = \varepsilon E$ that presents E as the cause and D as its effect, as was mentioned in Sec.1.5, the expression for the energy stored per unit volume in electric field in different forms is

$$SE_e = \frac{ED}{2} = \frac{\varepsilon E^2}{2} = \frac{D^2}{2\varepsilon} \qquad \text{.................................} \quad (1.48)$$

This generic field equation holds true for any point in an electric field of any configuration. In the case of uniform or parallel field, as in parallel plate capacitor, SE_e is the same at all points in the field. In non-uniform fields, like a radial field, it varies from point to point in the field.

As will be seen subsequently, there is very close analogy between Electrostatics, Magnetism, and Electromagnetism. In Electrostatics, we found both the Q- and ψ-concepts applicable. In Magnetism with Isolated Pole Theory, the corresponding m- and ϕ- concepts will be found applicable (Sec.3.14). In Electromagnetism, there is no isolated magnetic pole, and instead, an electric current produces the magnetic flux; so the ϕ- concept is the only option (Sec.4.7.1). However, an alternative I- concept will be found in Sec.5.4 of Electromagnetic Induction, which latter, in effect, is only an extension of Electromagnetism. These comments will be referred back to during the discussions in subsequent chapters.

1.15 Electromechanical Energy Conversion in Electrostatics:

In this section, we consider what happens if there be relative motion between the plates of a charged capacitor. Suppose the positive plate A of the parallel plate capacitor shown in Fig.1.37 is static and the negative plate B is allowed to move by the attractive force. Two situations may arise. The battery that charges the capacitor may remain disconnected during movement of plate B, so that the charges $\pm Q$ on the plates remain unaltered (Q– constant condition), or the battery may remain connected, so that the PD across the plates V remains unaltered (V– constant condition). We consider these situations separately.

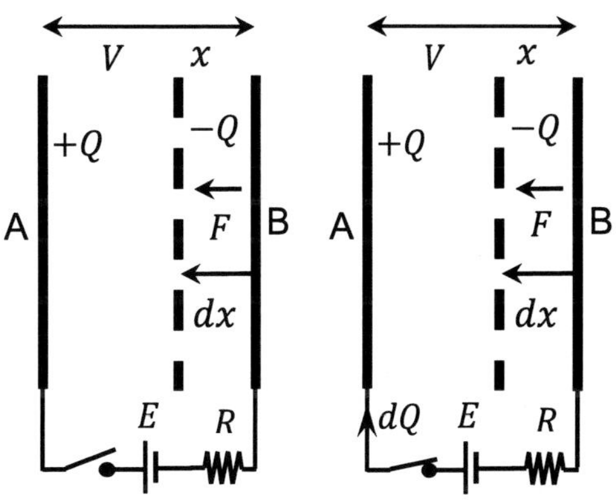

(a) Q-constant (b) V-constant

Fig.1.37 Electromechanical energy conversion in Electrostatics

(A) Q – constant condition: In Fig.1.37(a), the capacitor of capacitance C is first energised to charges $\pm Q$ and potential difference $V = Q/C$, and the battery is then disconnected, as shown. During subsequent movement of plate B, as the charge Q remains unaltered, field strength between the plates, $E = Q/(\varepsilon A)$, and the attractive force on plate B, $F = Q^2/(2\varepsilon A)$, remain constant. If F moves plate B by distance dx in time dt, $V = E.x$ decreases by $dV = E.dx$, and energy stored in the field, $SE_E = QV/2$, decreases by $d(SE_E) = Q.dV/2$. Thus we have,

Loss in PE_E of plate B = Work done by F or the mechanical output from the system

$$-d(PE_E) = F.dx = \frac{QV}{2x}dx \quad \text{(by Eqn.1.40)}$$

$$= \frac{QE.x}{2x}dx$$

$$= \frac{QE.dx}{2}$$

$$= \frac{Q.dV}{2}$$

$$= -d(SE_E)$$

$=$ Loss in SE_E in the electric field.

The interpretation of this result is that the stored energy in the electric field has been expended in delivering mechanical output in the form of loss in potential energy of plate B, complying to the relation

$$\frac{dSE_E}{dx} = \frac{dPE_E}{dx} \ (= -F) \text{ we obtained in Sec.1.14.}$$

(B) V – constant condition: In Fig.1.37(b), the battery remains connected to the capacitor during the force F moves plate B, as shown, so V remains unaltered. We have in general,

$$C = \varepsilon A/x$$

or, $$\frac{dC}{dx} = \varepsilon A \frac{d}{dx}\frac{1}{x} = -\frac{\varepsilon A}{x^2} = -\frac{C}{x}$$

where dx is positive, or an increment in x along positive direction. Hence, if dx is a decrement, we have

$$\frac{dx}{x} = \frac{dC}{C} \quad \text{.................................... (A)}$$

This identity also prevails in Q-constant condition, but we did not require using the same. As C increases by dC, $Q = CV$ increases by $dQ = V.dC$, and this dQ amount of charge is drawn from the battery in time dt, constituting a current $i = dQ/dt$. Voltage equation of the circuit is

$$E = iR + V$$

and the energy equation for time dt is

$$Eidt = i^2Rdt + Vidt = i^2Rdt + VdQ$$

or, Battery output = Heat in R + Energy added to the system

Thus, whereas the battery emf is E, the voltage across the capacitor is maintained at V during time dt, due to the iR drop. We have in this case

Loss in PE_E of plate B = Work done by F or the mechanical output from the system

$$-d(PE_E) = F.dx$$

$$= \frac{CV^2}{2x}dx \quad \text{(by Eqn.1.40)}$$

$$= \frac{V^2}{2}C\frac{dx}{x}$$

$$= \frac{V^2}{2}dC \quad \text{(by Eqn.A)}$$

$$= \frac{V.dQ}{2}$$

Also, the energy stored in the field, $SE_E = QV/2$, increases by $d(SE_E) = V.dQ/2$.
Interpretation of the result in this case is that out of the energy $V.dQ$ received from the battery (earning), half of this, $V.dQ/2$, has been stored in the electric field (saving), and the rest half has been utilised (expenditure) in delivering mechanical output in the form of loss in the potential energy of plate B.

Although the subject has been discussed with reference to parallel plate capacitor, the result obtained is of general validity.

This form of energy conversion is not practicable in electrical power engineering applications. One application in electrical instrumentation is the electrostatic voltmeter, wherein a variation of voltage to a spring-controlled movable-plate capacitor results in a change in the mechanical deflecting torque in the instrument.

Ch. 2: Current Electricity

2.1 Introduction: The history of current electricity can be traced back to 1789. Luigi Aloisio Galvani (1737-1798), an Italian professor of anatomy at Bologna, happened to notice that the muscles of fresh-skinned frog's legs twitched when touched by two dissimilar metal sticks. His followers attributed the cause to 'animal electricity'. Alessandro Giuseppe Antonio Anastasio Volta (1745-1827), an Italian physicist and a professor of experimental physics at the University of Pavia, was among the first scientists who repeated and checked Galvani's experiments. In 1793, he established that Galvani's animal electricity is just ordinary electricity, and he invented the first electric cell. He also developed the concept of 'completing an electric circuit'. In 1799 he devised an electric battery (combination of electric cells) which is known as 'Voltaic Pile'. In 1807, the famous British scientist Sir Humphrey Davy (1778-1829) established that the essential element of voltaic pile is chemical reaction. In 1821, he discovered that the current through a conductor experiences a 'resistance' and this resistance is proportional to the length of the conductor and inversely proportional to its cross sectional area. He also discovered that the resistance of a conductor increases with its temperature. In 1827, Georg Simon Ohm (1789 -1854), a German physicist and mathematician, established that the driver of electric current is potential difference, and discovered their relationship in the form of ohm's law. Unfortunately, Ohm's discovery was not recognised at that time. In 1841, after 14 years, Ohm's law was recognised by the Royal Society. In 1841, James Prescott Joule (1818-1889), FRS and English physicist, discovered the heating effect of electric current in the form Joule's first law $(H = i^2Rt)$. Gustav Robert Kirchhoff (1824-1887), a German physicist, formulated his circuit laws (current and voltage laws) in 1845, at the age of twenty one when he was still a student. So far as coverage in this book is concerned, the topics in Current Electricity are not of so much conceptual complexity compared to those in other chapters. As such, most topics have been kept out of scope, especially the circuit laws and theorems, including series parallel combination of resistances. Only some selected topics have been summarised in simple words to (a) cover common quantities and their units, and (b) supplement some topics in other chapters. Further, the topics have been discussed with reference to metallic conductors only.

2.2 Electric Source and its Electromotive Force: An electric source (*voltage source*, as distinct from *current source*) may be described as an agency capable to deliver electric charge, electric current, or electrical energy in general, continuously at constant potential difference developed within itself. This inherent PD of an electric source is called its *Electromotive Force*, abbreviated as EMF or emf. Emf is denoted by E or e. Being electric potential difference (discussed in Ch.1), the unit of emf remains the volt. A DC source delivers direct current and an AC source delivers alternating current. A DC source may be electric cell, DC generator, converter, etc. A combination of cells is called an electric battery. In a battery, a number of cells are connected in series, in parallel, or in both, to increase respectively the emf, the current delivering capacity, or both. An AC source may be AC generator (alternator), transformer secondary, inverter, etc.

2.3 Electric Current: Continuous flow of electric charge constitutes an electric current. It is denoted by the symbol I. If Q amount of charge crosses a conductor section in time t, the electric current through the conductor

$$I = Q/t \qquad \text{.......................... (2.1)}$$

Electric current is thus coulomb per second (C/s). It is named 'ampere' (abbreviated as Amp and denoted by A), after the famous French scientist Andre Marie Ampere.

In Fig.2.1, we consider a length l of a metallic conductor of uniform cross sectional area A. If n number of 'free' electrons are there per unit volume of the conductor material, total charge within the length l, $Q = neAl$, where e is the electronic charge, 1.601×10^{-19} C. If this amount of charge crosses any section like XX' of the conductor in time t, the current becomes

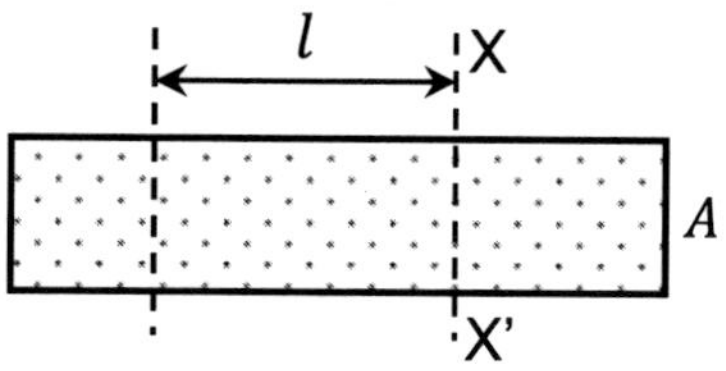

Fig.2.1 Electric current

$$\begin{aligned} I &= Q/t \\ &= neAl/t \\ &= neAv \end{aligned} \qquad \text{.......................... (2.2)}$$

where v is the drift velocity of the electrons. Actual drift velocity of electrons is very small, of the order of some metre per hour. However, as the electrons are densely packed in materials, the electric current is propagated at the speed of light. As electrons are negative charge carriers, the conventional direction of electric current is opposite to that of the electron flow. An electric current has four major effects. These are

(a) Thermal effect (electric heating)

(b) Lighting effect (illumination)

These are forms of Radiation effect.

(c) Magnetic effect (electromagnets), and

(d) Chemical effect (electrolysis).

2.4 Electrical Resistance – Ohm's Law: Metals are composed of lattice (crystal) structure of atoms. When an electric potential difference (voltage) is applied across a metallic conductor, its free electrons drift under the influence of the electric field. In the process, the electrons collide with the crystal lattices, and their mobility is impeded. This creates the *electrical resistance* of the conductor that restricts the flow of the current through it. Imperfections in the lattices and impurities in the metal also contribute significantly to its resistance. Electrical resistance is denoted by the symbol R. The current I through a conductor of resistance R, as driven by a potential difference V across the conductor, is given by Ohm's law as

$$I = V/R \qquad \text{.......................... (2.3)}$$

The unit of resistance is thus volt/ampere (V/A), and it is named 'ohm' (denoted by the Greek capital letter omega Ω) after the German physicist Georg Simon Ohm who discovered this proportionality law.

At any given temperature, the resistance of a conductor is decided by its geometry and material composition. If l be length of a conductor along the direction of current flow, and A its uniform cross sectional area, its resistance is given by

$$R = \rho l/A \qquad \text{.............................. (2.4)}$$

where ρ is a material constant called its *resistivity or specific resistance*. Eqn.2.4 asserts that the resistivity of a material is the resistance of a specimen of the material of unit length and unit cross sectional area.

Eqn.2.4 also asserts that dimensionally the resistivity is 'resistance x length' ($R \times l$), and so its unit is ohm-metre (Ω-m).

The reciprocal of resistance is called the *conductance*, denoted by G. Its unit is ampere/volt (A/V), and is named 'siemens' (S) after the German inventor and industrialist Ernst Werner Siemens (1816-1892) who was also the founder of the electrical and telecommunications company named Siemens. We have

$$G = \frac{1}{R} = \frac{1}{\rho}\frac{A}{l} = \gamma A/l \qquad \text{.............................. (2.5)}$$

Reciprocal of resistivity $1/\rho$ is called the *conductivity or specific conductance*, denoted by γ. Conductivity is thus a material property, and Eqn.2.5 asserts that the conductivity of a material is the conductance of a specimen of the material of unit length and unit cross sectional area. Eqn.2.5 also asserts that conductivity is dimensionally 'conductance / length' (G/l), and so its unit is siemens per metre (S/m). Whereas the resistivity of a material is a measure of its ability to restrict the flow of electricity through it, its conductivity is a measure of its ability to conduct electricity through it.

2.5 Temperature Coefficient of Resistance: As the temperature of a metallic conductor increases, its atomic vibrations increase. As such, the drifting electrons encounter more collisions with the crystal lattices, and the drift velocity decreases. The effect we get externally is a reduction in the current, or an increase in the resistance, for a given value of the PD.

If R_1 and R_2 be the resistances of a conductor at temperatures t_1 and t_2 respectively, their relationship is given by

$$R_2 = R_1\{1 + \alpha(t_2 - t_1)\}$$

$$\text{or, } R_2 - R_1 = \alpha R_1(t_2 - t_1)$$

where α is a property of the conductor material, called its temperature coefficient of resistance. Its unit, as can be derived from this relation, is per unit temperature difference, so is *per Celsius degree* $(C^0)^{-1}$. (It is practice to use *degree Celsius* (^0C) for temperature and *Celsius degree* (C^0) for temperature difference). The temperature coefficient of resistance of any material is itself temperature-dependent, and this temperature dependence for a metallic conductor can be derived from the following consideration. Resistance of metallic conductors varies linearly with temperature over a wide range of temperatures commonly encountered in practice, as shown by the line AB in Fig.2.2. The linearity, however, deviates at some fairly large negative temperature, as shown. If the linearity were maintained, as shown dotted, the resistance of the conductor would be zero at some temperature $-t_0$. This minus temperature t_0 is a characteristic of the conductor material, called its *inferred zero-resistance temperature*. From the triangles in the figure, we have

$$\frac{R_2}{R_1} = \frac{t_0+t_2}{t_0+t_1}$$

or, $$\frac{R_2-R_1}{R_1} = \frac{t_2-t_1}{t_0+t_1}$$

or, $$R_2 - R_1 = R_1\left[\frac{1}{t_0+t_1}\right](t_2 - t_1)$$

$$= \alpha R_1(t_2 - t_1)$$

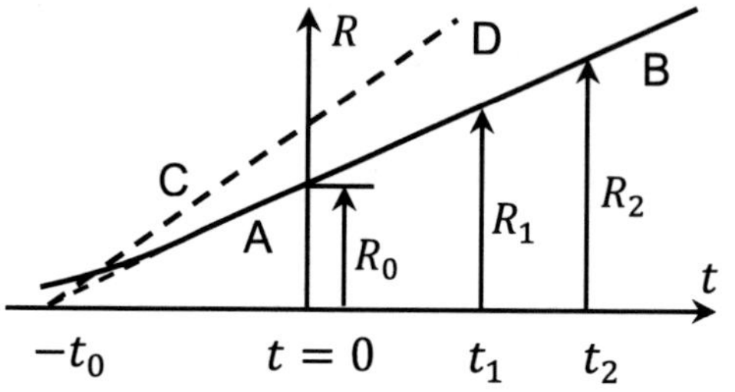

Fig.2.2 Temperature dependence of α for metallic conductor

Hence, $\alpha = \frac{1}{t_0+t_1}$ is dependent on t_1,

so we write $\alpha_1 = \frac{1}{t_0+t_1}$. Now, $\frac{R_2}{t_0+t_2} = \frac{R_1}{t_0+t_1} = \frac{R}{t_0+t}$ is constant for a given specimen of conductor, but not for its material. Line CD with stepper slope applies for a higher resistance of same material. It is thus explained why the temperature coefficient of resistance is itself temperature dependent, even for metallic conductors with linear resistance-temperature characteristics.

Based on the above discussion, the resistance temperature related equations are modified as

$$R_2 = R_1\{1 + \alpha_1(t_2 - t_1)\} \qquad \text{............................ (2.6)}$$

or, $R_2 - R_1 = \alpha_1 R_1(t_2 - t_1)$,

and $\alpha_1 = \frac{1}{t_0+t_1}$

It can be seen that $(R_2 - R_1)$ is proportional to R_1 and α_1.

Similarly, $\alpha_2 = \frac{1}{t_0+t_2}$ and $\alpha_0 = \frac{1}{t_0}$. It follows that $\alpha_1 = \frac{1}{(1/\alpha_0)+t_1}$ and $\alpha_2 = \frac{1}{(1/\alpha_0)+t_2}$.

Also, $\alpha_2 = \frac{1}{t_0+t_2} = \frac{1}{t_0+t_2+t_1-t_1} = \frac{1}{(t_2-t_1)+(t_0+t_1)} = \frac{1}{(t_2-t_1)+(1/\alpha_1)}$

These relations are compiled as

$$\left.\begin{aligned} \alpha_0 &= \frac{1}{t_0} \\ \alpha_1 &= \frac{1}{t_0+t_1} = \frac{1}{(1/\alpha_0)+t_1} \\ \alpha_2 &= \frac{1}{t_0+t_2} = \frac{1}{(1/\alpha_0)+t_2} = \frac{1}{(t_2-t_1)+(1/\alpha_1)} \end{aligned}\right\} \qquad \text{............ (2.7)}$$

As the average atmospheric temperature is nearly 20^{0}C, it is customary to express the temperature coefficient of resistance at this temperature as α_{20}, so also the resistivity at 20^{0}C as ρ_{20}. Table 2.1 shows these properties for few common conductor and resistor materials. A negative value of α implies decrease in resistance with increase in temperature. α is positive for metallic conductors, but negative for carbon, electrolytes, semi-conductors and insulating materials. It is almost zero for certain alloys over a very wide range of temperature.

Table 2.1: Conducting properties of some common metallic conductor and resistor materials

Material	Composition	ρ_{20} (ohm-m x 10^{-8})	α_{20} (per C^0)	$-t_0$ (^0C)
Conductor materials				
Silver (Ag)	Element	1.6	0.0038	310
Copper (Cu)	Element	1.7	0.004	234.5
Gold (Au)	Element	2.44	0.0034	274
Aluminium (Al)	Element	2.8	0.004	236
Zinc (Zn)	Element	6	0.004	230
Brass	Cu-Zn alloys	6 to 8	0.002	480
Nickel (Ni)	Element	7.8	0.006	147
Iron (Fe)	Element	12	0.0056	180
Lead (Pb)	Element	22	0.0039	236
Mercury (Hg)	Element	96	0.0009	1100
Resistor materials				
Tungsten (W)	Element	5.5	0.0045	200
Nichrome	Ni-Cr alloys	90 to120	0.00017	2250
Manganin	Cu alloys	42 to 74	0.00003	236
Constantan	Cu-Ni alloy	45 to 52	0.000008	125000

2.6 Electrical Energy and Power: If Q amount of charge moves or flows between two points in an electric circuit having a constant PD of V during time t constituting an electric current I, work done or energy expended by the electric source

$$W = VQ$$

And the time rate of energy is power

$$P = W/t = VQ/t = VI$$

In SI units, Q is in coulomb, t is in second, I is in ampere, and V is in volt. Electrical energy W is measured in the unit of joule (J), named after James Prescott Joule, which is the same as newton-metre (N.m), the unit of mechanical energy. Electrical power is measured in watt (W), named after the Scottish inventor and mechanical engineer James Watt (1736-1819) who is famous for inventing the condenser that resulted into a revolutionary improvement in the efficiency of steam engines. The unit watt is joule per second (J/s) or newton-metre per second (N.m/s). Hence, electrical energy is watt-second (W.s). In practice, it is convenient to use the units kilowatt (kW) for electrical power and kilowatt-hour (kWh) for electrical energy. We have, kWh = 10^3 W x (60x60) s = 3.6 x 10^6 W.s or J. In high voltage systems involving very large amount of energy at high power, kV is used for PD, MW for electrical power and MWh for electrical energy.

2.7 Heat and its Mechanical Equivalent: When a mass of material is heated, increased molecular vibrations within the mass increase its temperature (provided the material does not change phase or state). For any given mass m of any material, its increase in temperature Δt is proportional to the amount of heat added ΔH, and the relationship is expressed as

$$\Delta H = mc\Delta t$$

where c is a property of the material called its specific heat. The MKS unit of heat is kilocalorie (k.cal), and its size is defined by arbitrarily choosing the material as water and its specific heat as unity. Thus, one kilocalorie is the amount of heat needed to raise the temperature of one kilogram of water by one degree Celsius. However, the specific heat of any material is temperature dependent. Thus, the size of kilocalorie is refined as the amount of heat needed to raise the temperature of one kilogram of water by one degree from 14.5 to 15.5 degree Celsius. This is often referred to as 15^0 kilocalorie. A number of similar definitions of kilocalorie are there, but the above definition is popular in physics and engineering.

As heat is energy, it must have some equivalence to the mechanical or electrical energy. In 1845, Joule published a paper entitled "The Mechanical Equivalent of Heat". His experimental results concluded that if W joule of mechanical energy is converted into H kilocalorie of heat energy, the conversion factor called

$$\textit{Mechanical Equivalent of Heat}\ (\mathrm{J}) = \frac{W}{H} = 4186\ \mathrm{J/k.cal.}$$

The SI unit of any form of energy is joule, and this necessitates heat energy also be expressed in joule, instead of kilocalorie, using the above conversion factor. Similarly, the specific heat is expressed in joule per kg per Celsius degree (J/kg.C^0), instead of kilocalorie per kg per Celsius degree (k.cal/kg.C^0). Thus, specific heat of water at 15^0C, 1k.cal/kg.C^0 in MKS unit, is expressed as 4186 J/kg.C^0 in SI units. It may be seen that the numerical values of mechanical equivalent of heat and specific heat of water at 15^0C in SI units are the same, 4186.

2.8 Heating Effect of Electric Current: Heat is generated in a conductor (of any material) when electric current flows through it. This is referred to as heating effect of current. Electric heating originates from the drifting electrons colliding with the crystal lattices (as discussed earlier), which the resistance of the material is due to. It is, therefore, logical that the amount of heat increases with increase of current or of resistance. During electric heating, electrical energy W expended by the source of the current is converted into heat energy H in the conductor. In mathematical terms,

$$H\ (\text{energy, } W) = VQ = I^2Rt = V^2t/R \ \text{ joule} \qquad \ldots\ldots\ldots\ldots (2.8)$$

This relationship was first developed by Joule and is referred to as Joule's Law of Heating.

2.9 Emf and PD of an Electric Source: Referring to Fig.2.3, when an electric source of emf E and internal resistance r drives a current I through a load of resistance R, the load does not get the full emf of the source due to the voltage drop in the internal resistance of the source. The voltage or PD across the source reduces to $V = E - Ir$. This is also the voltage IR across the load terminals, neglecting the resistance of the leads connecting the source and the load. The load voltage, therefore, varies with load current, describing a term 'voltage regulation'. Lower the internal resistance of a source, higher is the load voltage, and better (lower) is the voltage regulation – a desirable condition in electrical power systems.

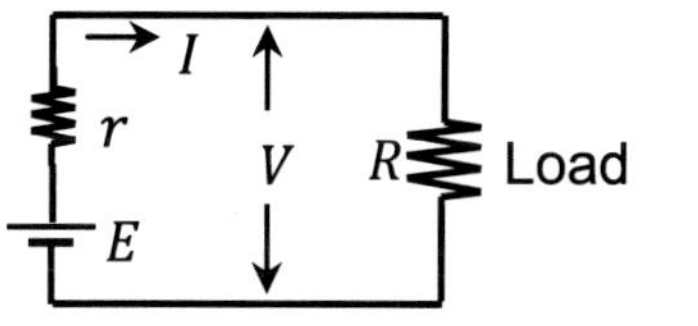

Fig.2.3 Emf and PD ($E - Ir = V = IR$)

2.10 Imperfect or Leaky Capacitor: As we know, a capacitance C exists between two neighbouring conductors A and B, say the two plates of a capacitor. The dielectric between the conductors being insulating material, it possesses some high insulating resistance r, and as such, a very small leakage current flows through a charged capacitor. Were the resistivity of the dielectric infinitely large, there would be no leakage current and the capacitor would be ideal or perfect. All practical capacitors are, therefore, imperfect or 'leaky' to more or less extent and this is the reason a charged capacitor loses its charge or gets discharged gradually with time, after the charging source is disconnected. A leaky capacitor is represented by the equivalent circuit shown in Fig.2.4(a).

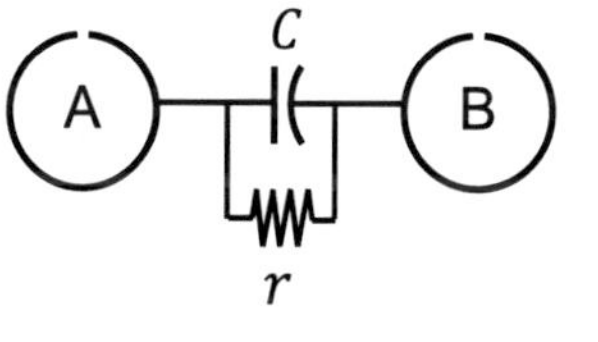

Fig.2.4(a) Equivalent circuit of leaky capacitor

The conductors A and B also have their individual insulation resistances r_1 and r_2 to earth, in addition to their earth capacitances C_1 and C_2 respectively. All these parameters need be considered when a capacitor is connected to a DC source, Fig.2.4(b). Apart from the leakage current through the capacitor, an earth leakage current i also flows through r_1 and r_2. We have,

$$V_1 = ir_1,\ V_2 = ir_2$$

$$\text{and } E = V_1 + V_2$$

Hence,

$$\frac{V_1}{V_2} = \frac{r_1}{r_2} \text{ or, } \frac{V_1}{V_1+V_2} = \frac{r_1}{r_1+r_2}$$

$$\text{or, } V_1 = E\frac{r_1}{r_1+r_2} \text{ and similarly, } V_2 = E\frac{r_2}{r_1+r_2}$$

Division of E into V_1 and V_2 is thus decided by the resistances r_1 and r_2, *not* by the capacitances C_1 and C_2. In the ideal case with infinite insulation resistances, however, the division of E into V_1 and V_2 is decided by the earth capacitances C_1 and C_2

$$V_1 = E\frac{C_2}{C_1+C_2} \text{ and } V_2 = E\frac{C_1}{C_1+C_2}$$

as was derived in Sec.1.10.5.

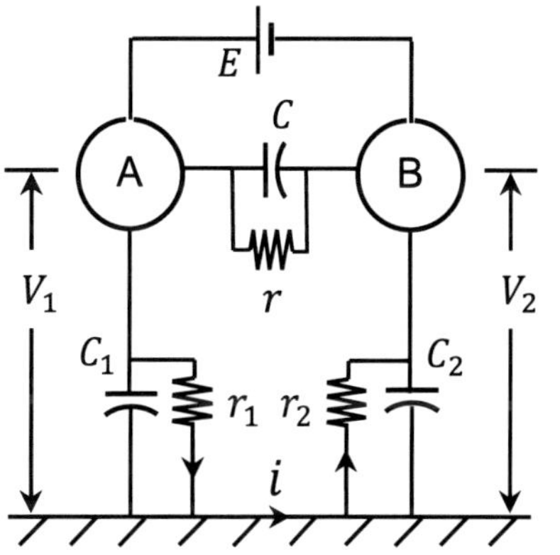

Fig.2.4(b) Voltage distribution in leaky capacitor

2.11 Charging and Discharging of a Capacitor:

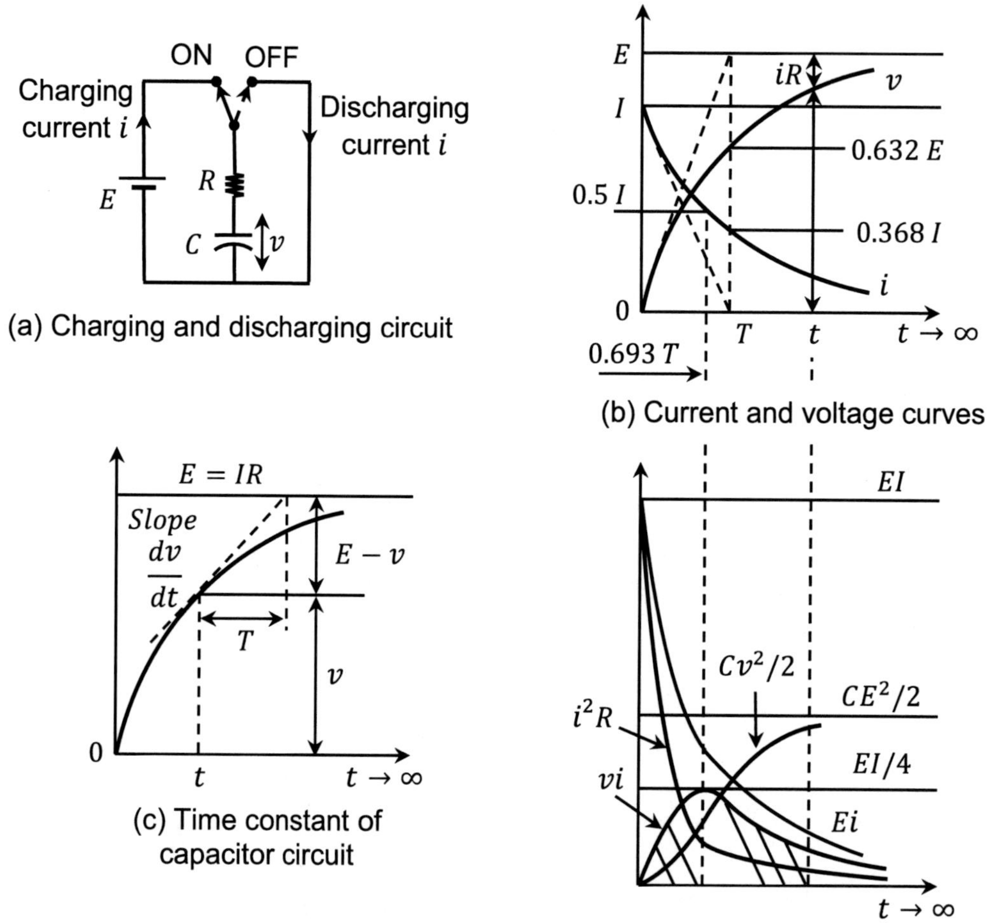

(a) Charging and discharging circuit

(b) Current and voltage curves

(c) Time constant of capacitor circuit

(d) Power and energy curves

Fig.2.5 Charging of capacitor

Referring to Fig.2.5(a), when a capacitor of capacitance C is switched ON to a DC source (say a battery), it is not charged to full source emf E instantaneously. As the charge q on the capacitor grows from zero, PD across its plates v builds up. This PD acts in opposition to the source emf, and as such, the charging current i gradually reduces from some high initial value to zero with time. At any instant of time t,

$$i = \frac{E-v}{R}$$

R being the charging circuit resistance. In next infinitesimal time period dt the charge on the capacitor grows by $dq = i.dt$, and its PD increases by $dv = dq/C$. Hence we have

$$i = dq/dt = C.dv/dt$$

Thus, $$\frac{E-v}{R} = C\frac{dv}{dt}$$

or, $$\frac{dt}{RC} = \frac{dv}{E-v}$$

Integrating both sides,

$$\frac{t}{RC} = -log_e(E - v) + A$$

where $e = 2.7183$ is the base of natural logarithm. This mathematical constant is also called as Euler's number or Napier's Constant, and should not be confused with emf. A is the constant of integration.

At the instant $t = 0$, $v = 0$, so $A = log_e E$. Therefore,

$$\frac{t}{RC} = -log_e(E - v) + log_e E = log_e \frac{E}{(E-v)}$$

$$\text{or, } \frac{E}{E-v} = e^{t/RC} \quad \text{or, } \frac{E-v}{E} = e^{-t/RC}$$

$$\text{or, } v = E(1 - e^{-t/RC})$$

$$= E(1 - e^{-t/T}) \qquad \text{........................ (2.9)}$$

where $T = RC$ (dimensionally time) is called the *time constant* of the circuit.

At the instant of switching $(t = 0)$, $v = 0$, $i = E/R = I$. This is the inrush charging current, as if the capacitor is not there in the circuit. As v grows with time, i falls off. At the instant t

$$i = \frac{E-v}{R} = \frac{E}{R} - \frac{v}{R} = \frac{E}{R} - \frac{1}{R}E(1 - e^{-t/T})$$

$$= I - I(1 - e^{-t/T})$$

$$= Ie^{-t/T} \qquad \text{.............................. (2.10)}$$

Thus, i first rises to I instantaneously, and then decays exponentially, as shown in Fig.2.5(b).

Significance of time constant can be appreciated from the following consideration. We had

$$\frac{dt}{RC} = \frac{dv}{E-v}$$

$$\text{or, } \frac{dv}{dt} = \frac{E-v}{RC} = \frac{E-v}{T}$$

Thus, referring to Fig.2.5(c), at any instant t during growth of the PD across capacitor plates, if the PD v continues to grow with the instantaneous rate dv/dt, the PD will attain its final value E in next time constant period T.

In Fig.2.5(b), the PD at the instant T $(t = T)$,

$$v = E(1 - e^{-1}) = E(1 - 0.368) = 0.632\,E$$

In words, the PD grows to 63.2% of its final value during initial time constant period. It can be checked that, the PD grows to 99.33% of its final value during initial $5T$ time period $(t = 5T)$ and to 99.99% of its final value during initial $10T$ time period.

Similarly, the charging current at the instant $t = T$

$$i = Ie^{-1} = 0.368\,I$$

That is, the current falls to 36.8% of its initial value during initial time constant period, to 0.673% during initial $5T$ time period, and to 0.004% during initial $10T$ time period.

Fig.2.5(d) explains the power and energy balance during the switching ON period of the circuit. Out of the total power Ei supplied by the source (battery) at any instant t, a portion is consumed as i^2R heating, and the rest vi contributes to the increase in the SE_E in the electric field, or $Ei = vi + i^2R$. The peculiar trend of vi curve requires an explanation. As $v + iR = E$ is constant, it is mathematics that the product $v.iR$ will exhibit a maximum value, or the product $v.i$ will exhibit a maximum value (R being constant), when $v = iR = 0.5E$. Hence the condition for this maximum value to occur

$$iR = IRe^{-t/T} = Ee^{-t/T} = 0.5E, \quad \text{or, } e^{-t/T} = 0.5 \quad \text{or, } e^{t/T} = 2,$$

$$\text{or, } t/T = log_e 2 = log_{10}2/log_{10}e = 0.3010/0.4343 = 0.693,$$

$$\text{or } t = 0.693\,T$$

This condition for maximum value may also be obtained using *maxima minima rule* of mathematics, as below.

$$f(t) = vi = E(1 - e^{-t/T}).Ie^{-t/T} = EIe^{-t/T} - EIe^{-2t/T} \quad \text{....... (A)}$$

$$f'(t) = EIe^{-t/T}(-1/T) - EIe^{-2t/T}(-2/T)$$

Equating $f'(t)$ to zero, we get $e^{-t/T} = 1/2$, and thus get the condition $t = 0.693\,T$. To confirm this condition for maximum value,

$$f''(t) = \frac{EI}{T^2}e^{-t/T} - \frac{4EI}{T^2}e^{-2t/T}$$

$$f''(c) = \frac{EI}{T^2}\frac{1}{2} - \frac{4EI}{T^2}\frac{1}{4} = -\frac{EI}{2T^2}$$

As $f''(c)$ is *negative*, $e^{-t/T} = 1/2$ gives the condition for *maximum* value of vi.

Putting $e^{-t/T} = 0.5$ in Eqn.(A) above, the maximum value of vi

$$(vi)_{max} = \frac{EI}{4} \quad \text{.. (2.11)}$$

and this occurs at

$$t = 0.693\,T \quad \text{.. (2.12)}$$

as shown in Fig.2.5(d). The shaded area below the vi curve upto instant t

$$\int_0^t vidt = C\int_0^t \frac{dv}{dt}vdt = C\int_0^v v\,dv = \frac{1}{2}Cv^2$$

gives the SE_E in the electric field upto the period the capacitor is charged to PD v. This energy graph is thus proportional to v^2, as shown. When the capacitor is charged full source emf E,

$$SE_M = \frac{1}{2}CE^2$$

as was discussed in Sec.1.14 in Electrostatics.

We now discuss the discharging of a charged capacitor. After the capacitor in Fig.2.5(a) is fully charged to the source emf E, if the switch is turned to discharging (OFF) side, the charges on the capacitor plates will be neutralised in the form of a current from the positive plate to the negative plate through the discharging link. The direction of the discharging current through the capacitor circuit is thus opposite to that of the charging current. Mathematical treatment of the discharging process follows.

At any instant of time t during discharging, let

v = PD remaining across the capacitor

q = Charge remaining across the capacitor, and

i = Instantaneous current.

We have, $i = \frac{v}{R}$. In next infinitesimal time period dt the charge on the capacitor reduces by $dq = i.dt$, and its PD reduces by $dv = dq/C$. Hence we have, $i = dq/dt = C.dv/dt$. Thus,

$$\frac{v}{R} = C\frac{dv}{dt} \text{ or, } \frac{dt}{RC} = \frac{dv}{v}$$

Integrating both sides,

$$\frac{t}{RC} = -log_e v + A$$

At the instant $t = 0$, $v = E$, so $A = log_e E$. Therefore,

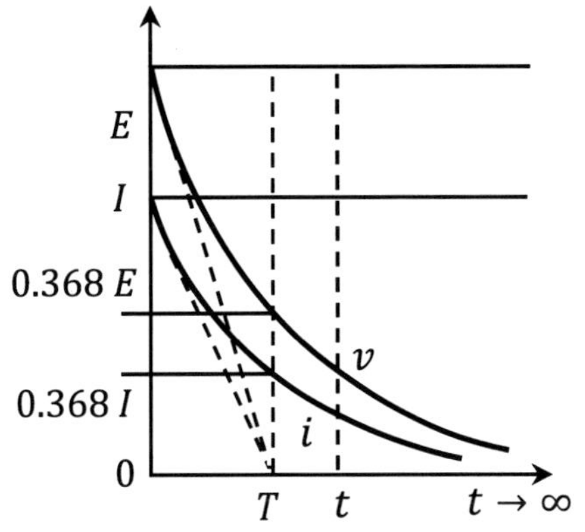

Fig.2.6 Discharging of capacitor

$$\frac{t}{RC} = -log_e v + log_e E = log_e \frac{E}{v}$$

$$\text{or, } \frac{E}{v} = e^{t/RC} \text{ or, } \frac{v}{E} = e^{-t/RC}$$

$$\text{or, } v = Ee^{-t/RC} = Ee^{-t/T} \quad \text{.......................... (2.13)}$$

At the instant $t = 0$, $v = E$, $i = E/R = I$. This is the inrush discharging current, often seen in the form of a spark. As v reduces with time, i falls off. At the instant t

$$i = \frac{v}{R} = \frac{E}{R}e^{-t/T}$$

$$= Ie^{-t/T} \quad \text{.......................... (2.14)}$$

Theoretically, it takes infinite time for the above growths and decays to their final values, but in practice these happen in seconds or even in a fraction of second.

Ch. 3: Magnetism

3.1 Introduction: Magnetism deals with the behavior and properties of permanent magnets, as distinct from electromagnets wherein magnetism is created by electric current, (Ch.4). A magnet may basically be conceived as a piece of some ferrous material (iron, nickel, cobalt and their alloys) which exhibits the property of

(a) attracting other pieces of ferrous materials around its two opposite ends, and
(b) aligning its two opposite ends along earth's north and south directions, if allowed to rotate freely.

The north-seeking end is called the North Pole or N pole of the magnet and the south-seeking pole as the South Pole or S pole. A force also exists between the poles of neighbouring magnets. Like or similar poles repel each other, whereas unlike or opposite poles attract. Depending on the application, magnets are manufactured in a variety of shapes and sizes.

Because a magnet (magnetic pole) can only exert a force on another magnet (magnetic pole), it follows that the earth is also a magnet. Further, because unlike poles attract and like poles repel, it also follows that the magnetic north pole of the earth N_m exists at its geographic south region S_g, Fig.3.1, located over Antarctica, the southernmost continent of the earth. Similarly, the magnetic south pole of the earth S_m exists at its geographic north region N_g, located over the middle of the Arctic ocean surrounded by the countries Russia, Alaska (USA), Canada, Greenland (Denmark), Iceland, Norway, Sweden and Finland. Geomagnetism, the subject that deals with the magnetic behavior of the earth, is beyond our scope of discussion.

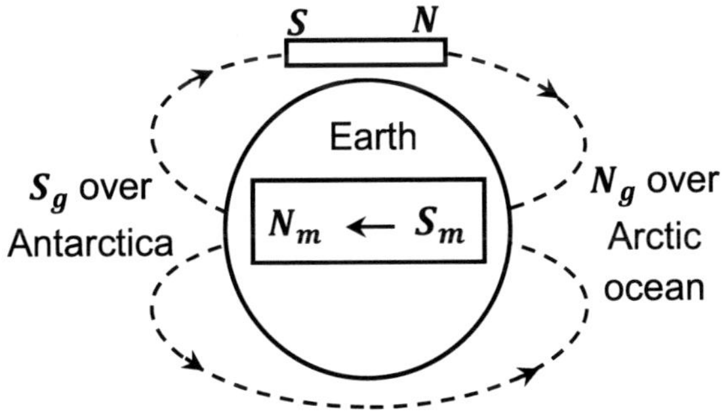

Fig. 3.1 The earth as a magnet

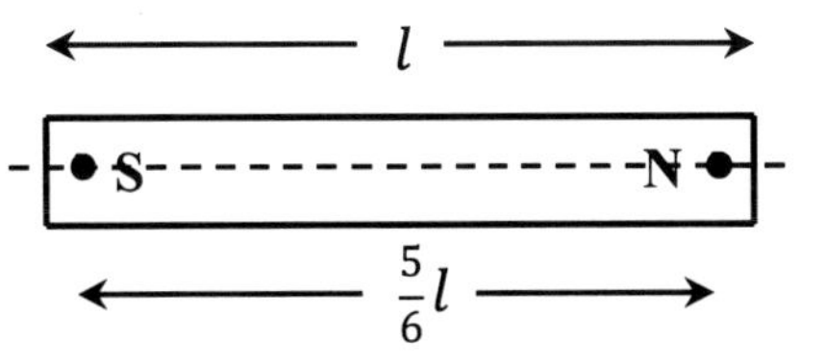

Fig.3.2 A bar magnet

Lodestone, Hercules Stone or Leading Stone, a mineral iron ore known today as Magnetite (Fe_3O_4), is a combination of ferrous oxide (FeO) or iron (II) oxide and another mineral iron ore called Hematite, which is ferric oxide (Fe_2O_3) or iron (III) oxide. Magnetite or lodestone is called iron (II,III) oxide, and it exhibits permanent magnetism due to certain combination of electron spins associated with the Fe^{2+} ions of FeO and Fe^{3+} ions in Fe_2O_3. Lodestone was first found in large quantities in Magnesia, in Asia Minor. It is learnt that in 900 BC, Magnus, a Greek shepherd, while walking across a field of loadstones, got the iron nails out of his sandals pulled by the stones. The pieces of lodestone are known today as first natural permanent magnets. The name *magnet* owes to its mother place of availability, Magnesia. In the twelfth Century, Chinese sailors started utilizing the site-seeking property of lodestone to navigate their ships.

A bar magnet is shown at Fig.3.2. The poles of a magnet are defined as the points where the force of attraction or repulsion is maximum. The poles of a magnet do not lie exactly at its ends but a little inside. The distance between the poles of a magnet is of the order of 5/6 th. of its length. For all practical purposes, however, the poles of a magnet are assumed to be uniformly distributed over its end surfaces. The line through the poles of a magnet is called its axis. A magnet thus consists of a pair of opposite poles separated by a distance, and as such, a single magnetic pole cannot physically exist in isolation. However, the basic laws of magnetism were developed with the concept of imaginary or fictitious isolated magnetic poles. Much later, after the atomic theory of matter was discovered & established, the origin of magnetism was explained in terms of the electric current rings produced by the rotating electrons in the atoms of matter. The *current-ring theory of magnetism*, however, reaffirmed the same laws of magnetism developed earlier with the concept of isolated pole. The current-ring theory thus superseded the isolated pole theory, but in view of the simplicity that the isolated pole theory offers in the understanding and derivation of the basic laws of magnetism, the *isolated pole theory* could not be discarded altogether, at least for introductory purpose. Another argument in favour of the isolated pole concept is that if imaginary electric or magnetic lines of force or flux is acceptable, why not the imaginary isolated pole concept. In this chapter, the basic laws of magnetism have been developed with the isolated pole theory, and equivalence between electromagnets and permanent magnets has been developed in Sec.4.8. Magnetic laws and quantities derived from the isolated pole concept are exactly analogous to their corresponding electrostatic counterparts, and in fact, this is the simplicity that greatly helps the beginner to grasp the magnetic theory easily in terms of the isolated pole concept. At many places in this chapter, therefore, the same language as in the corresponding places in Electrostatics has been intentionally used, just to feel the essence of analogy between Electrostatics and Magnetism. Hence, elaboration of some analogous topics in this chapter has been either skipped or simplified. Analogous electrostatic and magnetic quantities have been compared in Table 3.1 at the end of this chapter.

3.2 Magnetic Pole Strength and Permeability: Magnetic pole strength in Magnetism is analogous to electric charge in Electrostatics. It is denoted by the symbol m and is expressed in the unit weber (Wb). The N pole is arbitrarily considered positive and the S pole as negative. In 1820, the force between magnetic poles was found to obey the same Coulomb's Law as was established experimentally by Coulomb (1738-1806) for the electric charges in 1785.

Referring to Fig.3.3(a), Coulombs Law in Magnetism is stated by the mathematical expression

$$F = \frac{m_1 m_2}{4\pi\mu_0\mu_r x^2} \quad \text{.. (3.1)}$$

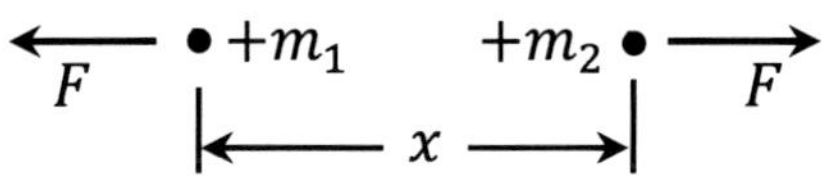

Fig.3.3(a) Coulomb's Law

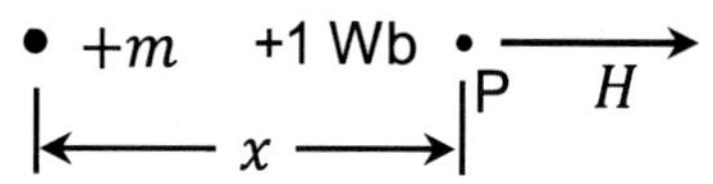

Fig.3.3(b) Field near point pole

where, F = Force in newton between two 'isolated' magnetic poles m_1 and m_2 weber

x = Distance between the poles in metre

μ_0 = Absolute permeability of free space (vacuum) which is also referred to as the *primary magnetic constant*, having a value of $4\pi.10^{-7}$ = 1.257 $\times 10^{-6}$ henry/metre (H/m). A note on the value of μ_0 is given in Sec.1.2. It will be seen later (Ch.5) that dimensionally inductance is given by permeability x length, and the unit of inductance being henry (H), permeability is expressed in the unit of henry/metre.

μ_r = Relative permeability of the medium (relative to the free space value) surrounding the magnetic poles. Being relative quantity, it is a unit-less number. Its value for free space is thus 1 by definition, and air may be taken to have the same value in place of its accurate value of 1.00000037. Unless mentioned otherwise, the medium remains understood as being the air, same as in Electrostatics.

The product $\mu_0.\mu_r = \mu$ is called the (absolute) permeability of the medium. Permeability in Magnetism is analogous to permittivity in Electrostatics. Permeability is the prime property of a magnetic material, and this magnetically classifies materials. The simplest classification, sufficient to understand the basic theory, is *Non-magnetic* and *Magnetic*. Non-magnetic materials cannot be magnetised to any appreciable extent, whereas magnetic materials can be easily magnetised. Non-magnetic materials have μ_r slightly more or less 1, and they do not exhibit hysteresis property (Ch.6). Magnetic materials are mostly ferrous group materials (iron, nickel, cobalt, and their alloys), although some ferrites and alloys of rare earth group elements are later developments as excellent permanent magnet materials (Table 6.1). Relative permeability of usual magnetic materials varies widely in the range from hundreds to several thousands, even in lakhs for some special materials, and they invariably exhibit hysteresis property. Whereas most dielectrics have single-figure ε_r values, most magnetic materials have 3- to 5-figure μ_r values. These are major distinctions in the properties of dielectric and magnetic materials. We generally use the term '*air*' to mean non-magnetic materials, and the term '*iron*' to mean magnetic materials. It follows from Eqn.3.1 that *the force between two unit magnetic poles (one weber each) situated one metre apart in vacuum is 63.325 $\times 10^{3}$ newton.* This defines the unit size of the magnetic pole, the weber.

3.3 Magnetic Field Strength: Magnetic field strength is analogous to electric field strength. In the space surrounding a magnet (or a system of magnets) some magnetic effect is experienced, and the space is said to be permeated with *magnetic field*. In other words, a magnetic field exists surrounding any system of magnets. *The magnetic field strength, magnetic field intensity, magnetic field or magnetising field* (the term *magnetising force* is now obsolete) H *at a point in a magnetic field is defined as the force on test pole placed at that point.* In similarity with the test charge in Electrostatics (Sec.1.3), a magnetic 'test pole' is defined as a unit point north pole without any magnetic field of its own to modify the mother field in which it lies. Like electric field strength, magnetic field strength is a vector quantity. As the field strength is force on unit pole, the force on a pole m lying in a magnetic field at a point of field strength H becomes

$$F = Hm \qquad \text{.......... (3.2)}$$

A negative value of m (south pole) makes F negative. This simply implies that the direction of the force F is opposite to that of the field H. Hence the sign of F may be ignored as far as its magnitude is concerned. The usual unit of magnetic field strength H is ampere/metre (A/m) which is the same as newton/weber (N/Wb), as is given by Eqn.3.2. It will be seen later (Sec.3.8) that dimensionally magnetic potential (or potential difference) is given by (magnetic field strength x length), and the unit of magnetic potential being ampere, magnetic field strength is expressed in the unit of ampere/metre. Referring to Fig.3.3(b), the magnetic field strength at a point distant x from and due to a point pole of strength m is given by

$$H = \frac{m}{4\pi\mu x^2} \qquad \text{.......... (3.3)}$$

3.4 Magnetic Flux: The concept of magnetic lines of force and magnetic flux is exactly similar to the electric lines of force and electric flux described in Sec.1.4. The line traced out by a test pole, if allowed to move freely in a magnetic field, is called 'a line of magnetic force'. Such lines of force thus originate from north pole and terminate on south pole, and the direction of a line of force through any point in a magnetic field is the direction of the field at that point. In any magnetic field there is infinite number of such lines of force, and the profile or distribution pattern of these lines of force is referred to as the configuration of the field. The distribution pattern of the lines of force due to a bar magnet, neglecting the geomagnetic field, is shown in Fig.3.14(b). Geomagnetic fields being much weaker than the practical magnetic fields commonly encountered in electrical machineries, the impact of the former on the latter is usually insignificant. Precision electrical and magnetic instruments, however, require magnetic shielding to safeguard against geomagnetic and other stray magnetic fields.

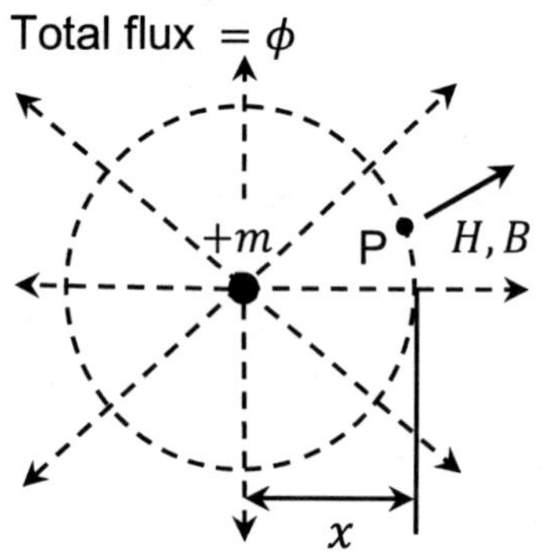

Fig.3.4 Flux and flux density around isolated magnetic pole

As shown in Fig.3.4, the lines of force around an isolated single pole are distributed radially in all directions in the entire three-dimensional space surrounding the pole. In line with the discussions in Electrostatics, magnetic lines of force is qualitative aspect of magnetic field, and '*magnetic flux*' is its quantitative aspect. In RMKS system of units or SI units, unit magnetic flux is arbitrarily assigned to be associated with unit magnetic pole, that is to say, unit magnetic flux originates from unit north pole and terminates on unit south pole. As such, magnetic flux is expressed in the same unit of magnetic pole, the weber, and it is denoted by the symbol ϕ. Thus, if ϕ weber of magnetic flux originates from or terminates on m weber of magnetic pole, we have by arbitration

$$\phi = m \qquad \ldots\ldots\ldots\ldots (3.4)$$

Like magnetic pole m, magnetic flux ϕ is a scalar quantity. It follows that the amount of magnetic flux associated with any given amount of magnetic pole is independent of the medium, or remains the same in all media, similar to electric flux. One distinction between the electric flux and magnetic flux must be noted. As electric charges can exist in isolation, electric flux lines, originating from positive charge and terminating on negative charge, may not form closed loops. On the other hand, magnetic poles being inseparable, magnetic flux lines, as shown in Fig. 3.14(b), continue through the magnet body and so always form closed loops.

3.5 Magnetic Flux Density: Magnetic flux density is analogous to electric flux density. *Magnetic flux density B at a point in a magnetic field is defined as the amount of magnetic flux crossing per unit area surrounding the point, the area being perpendicular or normal to the direction of the flux.* Thus, if ϕ weber of magnetic flux crosses a normal area A uniformly, the magnetic flux density at any point on the area is given by

$$B = \frac{\phi}{A} \qquad \ldots\ldots\ldots\ldots (3.5)$$

This is the general expression of magnetic flux density. It is expressed in weber per square metre (Wb/m^2). It is a vector quantity, the direction being along the positive direction of the flux ϕ normal to the area A. In SI units, Wb/m^2 has been renamed as tesla (T).

Referring to Fig.3.4, let us consider an imaginary sphere of radius x surrounding a point north pole (of strength) m at its centre. The total flux $\phi = m$ radiates normally and uniformly over the total surface area $A = 4\pi r^2$. Hence the flux density at any point P (on the surface of the sphere) distant x from and due to a point pole m is given by

$$B = \frac{\phi}{A} = \frac{m}{4\pi x^2}$$

Again from Eqn.3.3, the field strength at the point P, $H = \frac{m}{4\pi\mu x^2}$. Hence we have

$$B = \mu H \qquad \ldots\ldots\ldots\ldots (3.6)$$

This is a general field equation that holds true for any point in any configuration of the field. This equation shows that B and H differ by μ times, which is justified because field strength is dependent on the medium but the flux and the flux density do not. Furthermore, H is the cause and B is its effect, and they have the same direction.

3.6 Intensity of Magnetisation: Intensity of magnetisation plays dual roles, analogous to electric charge density (Sec.1.6) and electric polarization (Sec.1.11). Here we discuss the first role. *Intensity of magnetisation of a magnet is defined as its pole strength per unit cross sectional area.* It is denoted by J, or sometimes by I also. Use of the symbol I for intensity of magnetisation is discouraged, as I is the most usual symbol for electric current. Hence, for a bar magnet of pole strength m, as usually assumed distributed uniformly over its uniform cross sectional area A, its intensity of magnetisation is given by

$$J = \frac{m}{A} \qquad \text{..........} (3.7)$$

Defined as such, J is a scalar quantity, and is expressed in weber per square metre (Wb/m^2). Like electric polarization, intensity of magnetisation may also be defined as flux density vector (Sec.3.11), and when so defined it can also be expressed in the unit tesla (T) which is specific to flux density only.

3.7 Some Common Field Derivations:

3.7.1 Field between two oppositely magnetised parallel surfaces: Fig.3.5 shows two closely spaced parallel magnetised surfaces of opposite polarity, say the opposite faces of a closed axis magnet with an air gap. The N and S poles of equal strength m are assumed to be distributed over the pole faces of area A uniformly, with intensity of magnetisation $J = m/A$. If the pole faces are of circular section of radius R, and distant d apart, as shown, following similar derivation as in Sec.1.7.2 in Electrostatics, magnetic field (force on a test pole) in the air gap at point P lying on the magnetic axis distant x from the N surface, due to the pole m on the N surface alone

$$\frac{J}{2\mu}[1 - Cos\,\theta] = \frac{J}{2\mu}\left[1 - \frac{x}{\sqrt{x^2+R^2}}\right]$$

This force is repulsive one. An attractive force

$$\frac{J}{2\mu}\left[1 - \frac{d-x}{\sqrt{(d-x)^2+R^2}}\right]$$

is also exerted by the S surface on the test pole. The total field at P due to both the surfaces, therefore,

$$H = \frac{J}{2\mu}\left[2 - \frac{x}{\sqrt{x^2+R^2}} - \frac{d-x}{\sqrt{(d-x)^2+R^2}}\right]$$

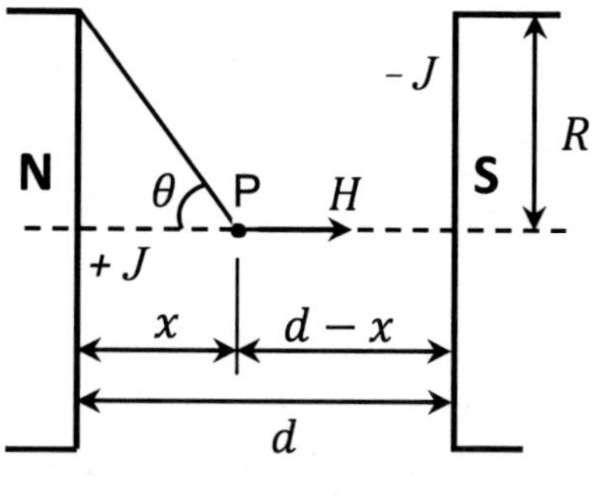

Fig.3.5 Field between two oppositely magnetised parallel surfaces

When the extent of the poles is very great compared with their distant apart, or $R \gg d$, the term within the square brackets approaches 2, and then

$$H = \frac{J}{\mu} = \frac{m}{\mu A} \qquad \text{..........} (3.8)$$

As was explained in Electrostatics, the condition $R \gg d$ implies that the "fringing effect" of the field in the air gap is negligible, so the field in the air gap is parallel or uniform. It is evident that when Eqn.3.8, that does not contain the term x, applies,

(a) the geometry of the pole faces does not bear any significance, that is, need not necessarily be circular but may be of any other shape, and as such the condition $R \gg d$ assumes the general form $A \gg d$, and also that

(b) the location of the point P may be anywhere between the poles, not necessarily on the axis. This is obvious, the field being uniform.

Same as in Electrostatics, the summary of this section is that *when $A \gg d$ in Fig.3.5, field at any point between the two surfaces due to the magnetic pole on any one surface is $m/(2\mu A)$, and so the field due to both the surfaces is $m/(\mu A)$, which is Eqn.3.8.*

Magnetic flux density at any point between and due to the two surfaces, when $A \gg d$,

$$B = \mu H = J = \frac{m}{A} \qquad \text{(3.9)}$$

3.7.2 Field on the axis of bar magnet:

Fig.3.6 shows a bar magnet of intensity of magnetisation J and length l. Field H at point P distant x from the centre of the magnet and on its axis is given by the resultant of the individual fields due to the N and S surfaces.

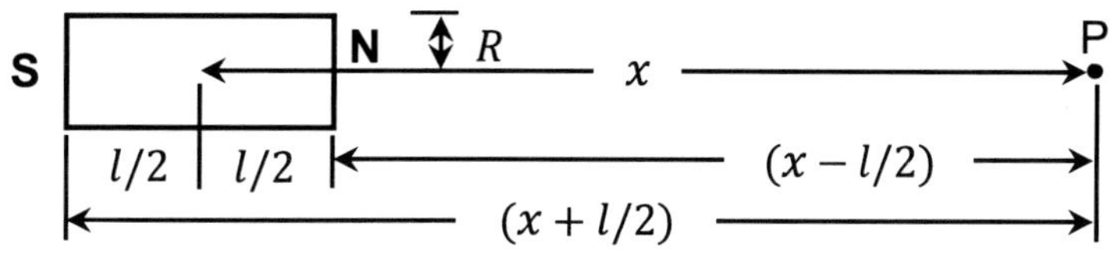

Fig.3.6 Magnetic field on the axis of bar magnet

Thus, if the magnet be cylindrical, of radius R, it follows from the previous section (3.7.1)

$$H = \frac{J}{2\mu}\left[1 - \frac{x-(l/2)}{\sqrt{[x-(l/2)]^2+R^2}}\right] - \left[1 - \frac{x+(l/2)}{\sqrt{[x+(l/2)]^2+R^2}}\right]$$

$$= \frac{J}{2\mu}\left[\frac{x+(l/2)}{\sqrt{[x+(l/2)]^2+R^2}} - \frac{x-(l/2)}{\sqrt{[x-(l/2)]^2+R^2}}\right] \qquad \text{(3.10)}$$

3.7.3 Field near short magnetic dipole:

Magnetic dipole resembles electric dipole, Sec.1.7.5. Thus, two equal and opposite point poles separated by a distance constitute a '*magnetic dipole*'. Following similar derivation as in Sec.1.7.5, the field strength H in Fig.3.7 at some point P distant x from the centre of a dipole of pole m and length l, when $x \gg l$, is approximately given by

Fig.3.7 Field near short magnetic dipole

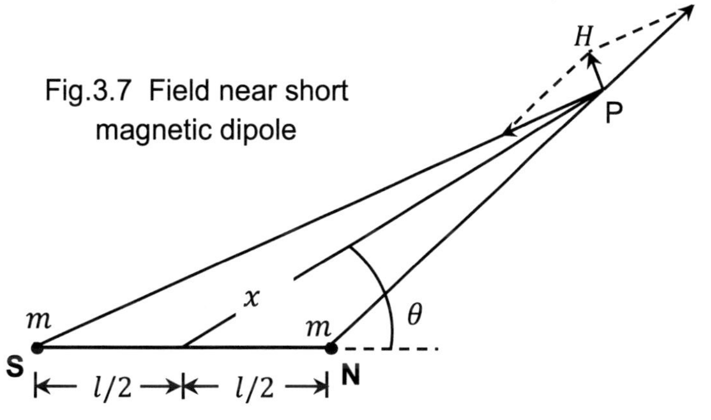

$$H = \frac{ml}{2\pi\mu}\frac{Cos\theta}{x^3} \qquad \text{(3.11(a))}$$

This equation has the same limitation as the corresponding equation 1.17(a) in Sec.1.7.5. For points on the axis of the dipole, $\theta = 0,\ Cos\,\theta = 1$, and H becomes

$$H = \frac{ml}{2\pi\mu x^3} \qquad \text{...} (3.11)$$

This is correct. As θ increases from zero, the degree of inaccuracy in Eqn.3.11(a) gradually increases, and finally on the perpendicular bisector line of the dipole ($\theta = 90^0$) the equation becomes invalid. It is noticed that the field strength near a magnetic dipole varies as $(1/x^3)$, whereas that near a point magnetic pole as $(1/x^2)$.

3.8 Magnetic Potential, Potential Difference and Potential Energy: Magnetic potential and potential difference (PD) are exactly analogous to electric potential and PD (Sec.1.8). Like electric potential as work done in electric field, magnetic potential is work done in magnetic field. Magnetic potential is a measure of magnetic potential energy, and is denoted by V_M.

Magnetic potential at a point in a magnetic field is defined as the amount of work done by some external agency in moving a test pole against the field from some arbitrarily chosen zero reference point to the specified point. The zero reference point has been discussed in Sec.1.8.

$$V_M = -\int H.dx \qquad \text{.........................} (3.12)$$

with limits of x along the direction of $(-H)$. Potential of a point may also be negative with respect to some zero reference point. Eqn.3.12 is in integral form; its differential form is $dV_M = -H.dx$, or $H = -dV_M/dx$. This simply implies that if H acts along positive direction of x, V_M decreases with increase in x along positive direction.

Magnetic potential difference (PD) between two points 2 and 1 in a magnetic field, denoted by $(V_M)_{21}$, *is defined as the amount of work done against the field in moving a test pole from point 1 to point 2.* Thus, $(V_M)_{21}$ denotes the potential of point 2 with respect to, or relative to, that of point 1. PD is, therefore, *relative potential*, as distinct from the *absolute potential* with respect to stated or implied zero reference.

$$(V_M)_{21} = -\int_1^2 H.dx \qquad \text{...} (3.13)$$

This is the general expression of PD. It follows that

$$(V_M)_{21} = -\int_1^2 H.dx = \int_2^1 H.dx = -\,(V_M)_{12}.$$

Hence, like potential the PD may also be positive or negative.

As we know (Sec.2.2), the inherent electric potential difference created within an electric source is called its electromotive force E. It will be observed in Ch.4 that the source of magnetic potential difference is electric current, and the magnetic potential difference created by an electric current is called the *magneto-motive force* (mmf), denoted by F, of the current. Like electric potential, PD or emf, the magnetic potential, PD or mmf is a scalar quantity. Electromagnetic considerations, Ch.4, reveal that the unit of magnetic potential difference is ampere (A), which is thus joule/weber (J/Wb), like the electric potential difference, the volt, being the same as joule/coulomb (J/C). The similarities between the corresponding electric and magnetic quantities and their units are thus appreciated gradually.

Magnetic potential energy of an isolated magnetic pole PE_M in a magnetic field is analogous to electrostatic potential energy of a charge PE_E in an electric field, Sec.1.8. *The amount of work done by some external agency in moving a magnetic pole against a magnetic field from some zero reference point of potential is stored with the pole as its magnetic potential energy* PE_M. It involves two independent agencies, a magnetic pole and a magnetic field, and must be expressed with respect to some stated or implied reference point. If a pole m lies in a magnetic field at some point of magnetic potential V_M (*created by some other pole or system of poles*), the magnetic potential energy of the pole m at that point, with respect to the stated or implied zero reference point of potential, becomes

$$PE_M = m.V_M \quad \text{...} \quad (3.14)$$

Hence, it may be positive or negative depending on the signs of m and V_M both. If V_M be the PD $(V_M)_{21}$ in Eqn.3.14, then the product $m.(V_M)_{21}$ gives the magnetic potential energy $(PE_M)_{21}$ of pole m at point 2 with respect to zero reference at point 1. If a pole m moves against a magnetic potential difference V_M, its magnetic potential energy PE_M increases by the amount $\Delta PE_M = Q.V_M$. Magnetic potential, PD and potential energy are all scalar quantities.

3.8.1 Magnetic potential and potential energy in the field of point pole: The derivation follows the same line as in the corresponding electric case, Sec.1.8.1. Thus, referring to Fig.3.8(a), magnetic potential of a point P distant x from and due to a point north pole m is computed as

$$V_M = -\int_{\infty}^{x} H.dx$$
$$= -\frac{m}{4\pi\mu}\int_{\infty}^{x}\frac{1}{x^2}.dx$$
$$= -\frac{m}{4\pi\mu}\left[-\frac{1}{x}\right]_{x=\infty}^{x=x}$$
$$= -\frac{m}{4\pi\mu}\left[-\frac{1}{x}\right]$$
$$= \frac{m}{4\pi\mu x} \quad \text{...............} \quad (3.15)$$

This is positive and is given by the positive shaded area in Fig.(a), also by the height 12 in Fig.(b). These positive curves in Figs.(a) and (b) also represent respectively the force F on, and the PE_M of, some north point pole at point P. The negative area and curves apply for some south pole at point P.

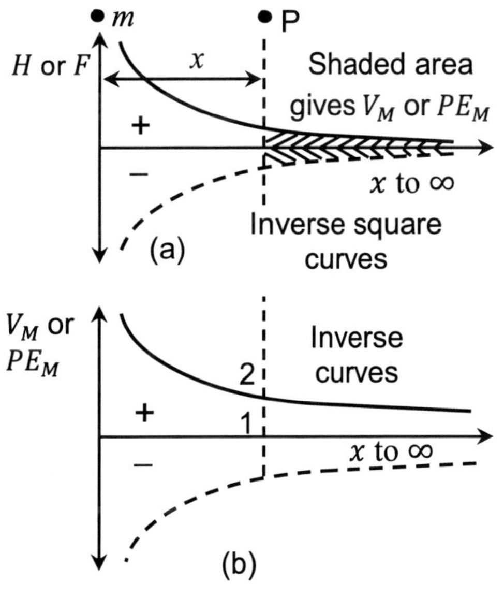

Fig.3.8 F, H, V_M and PE_M due to point pole

Like $F = -dPE_E/dx$ in Electrostatics, here we have $F = -dPE_M/dx$ with similar interpretation.

3.8.2 Magnetic potential and potential energy in the field of single magnetised surface: The derivations follow the same line as in the corresponding electric case, Sec.1.8.4. Thus, in Fig.3.9(a), AB is a circular surface of radius R, with right face magnetised to intensity of magnetisation $+J$ (north polarity), say the north pole face of a cylindrical magnet. P is a point distant x from the centre of the surface on its axis, as shown.

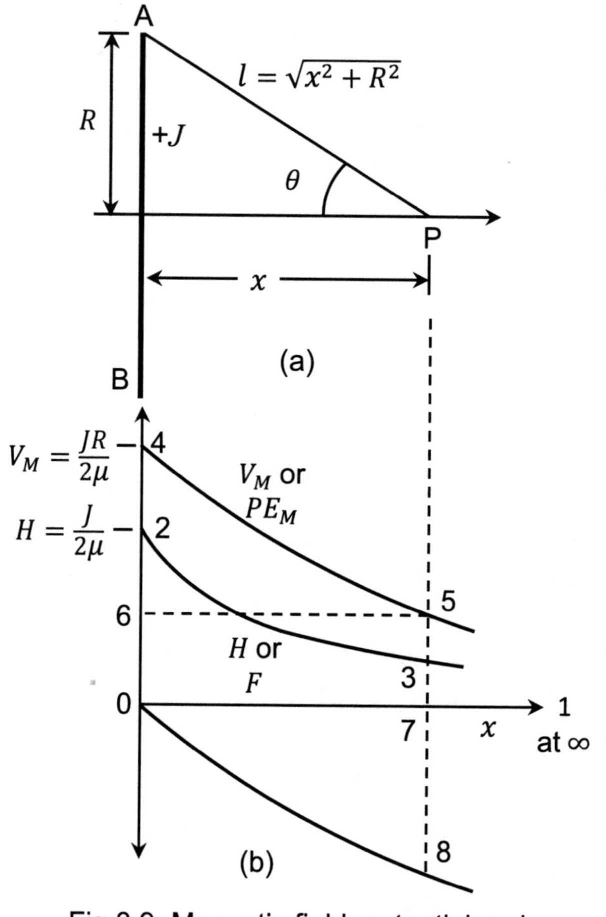

Fig.3.9 Magnetic field, potential and PE_M near single magnetised surface

Field strength at point P, as was derived in Sec.3.7.1,

$$H = \frac{J}{2\mu}(1 - Cos\theta)$$

$$= \frac{J}{2\mu}\left[1 - \frac{x}{\sqrt{x^2+R^2}}\right]$$

The H-x variation has been shown by the curve 2-3-1 in Fig.3.9(b). The same curve also represents, to different scale, the force F on some north pole at point P. Following similar derivation as in Sec.1.8.4, potential at point P

$$V_M = \frac{JR}{2\mu}tan\frac{\theta}{2} = \frac{J}{2\mu}\left[\sqrt{x^2 + R^2} - x\right] \quad \text{......................} (3.16)$$

Curve 4-5-1 shows the V_M-x variation. This potential is with respect to zero reference at $x = \infty$. Curve 08 is the plot of the potential with respect to zero reference at $x = 0$, that is, with respect to the surface AB. Thus, the potential V_M of point P with respect to zero reference at $x = \infty$ is the height 75, which is positive, and that with respect to the surface AB is the depth 46 or 78, which is negative. The same potential curves also represent, to different scale, the PE_M of any north pole at the point P.

3.8.3 Magnetic PD between two oppositely magnetised parallel surfaces: Magnetic potential difference V_M across the narrow air gap of a magnet with closed axis, Fig.3.5, similar to the electric potential difference between the two plates of a parallel plate capacitor derived in Sec.1.8.5,

$$V_M = H.d = \frac{m.d}{\mu A} = \frac{J.d}{\mu} \quad \text{............................} (3.17)$$

3.8.4 Magnetic potential on the axis of bar magnet: With reference to Fig.3.6, potential at point P on the axis of the cylindrical bar magnet is derived as below. Potential at P due to the N and S surfaces individually, by Eqn.3.16, are respectively

$$\frac{J}{2\mu}\left[\sqrt{[x-(l/2)]^2+R^2}-[x-(l/2)]\right] \text{ and } -\frac{J}{2\mu}\left[\sqrt{[x+(l/2)]^2+R^2}-[x+(l/2)]\right]$$

Hence the total potential at P due to both the surfaces

$$V_M=\frac{J}{2\mu}\left[\sqrt{[x-(l/2)]^2+R^2}-\sqrt{[x+(l/2)]^2+R^2}+l)\right] \quad \text{...... (3.18)}$$

3.8.5 Potential near short magnetic dipole: Magnetic potential near a short magnetic dipole can be derived in the same way electric potential near a short electric dipole was derived in Sec.1.8.6. Thus, referring to Fig.3.7, magnetic potential V_M at some point P distant x from the centre of a magnetic dipole of pole m and length l, when $x \gg l$, is given by

$$V_M=\frac{mlCos\theta}{4\pi\mu x^2} \quad \text{.. (3.19)}$$

This potential is with respect to zero reference at infinite distance. Further, magnetic potential near a dipole varies as $(1/x^2)$, whereas that near a point pole varies as $(1/x)$.

3.9 Magnetic Moment of Magnet: Magnetic moment or magnetic dipole moment in Magnetism is analogous to electric dipole moment in Electrostatics, Sec.1.9. Magnetic moment of a magnet specifies the strength of the magnet. For a bar magnet of pole strength m and length l, it is defined as a vector denoted by

$$\boldsymbol{p_m}=m.\boldsymbol{l} \quad \text{.. (3.20)}$$

directed from the S pole to the N pole, as shown at Fig.3.10(a). The subscript m denotes 'magnetic', and is often omitted.

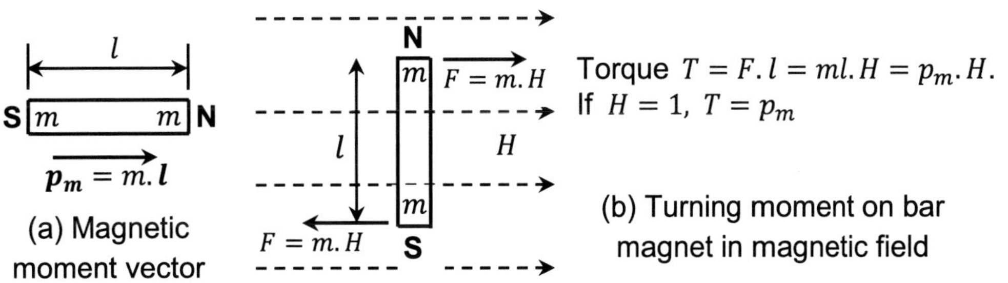

Fig.3.10 Magnetic moment explained

Significance of this definition can be appreciated from similar considerations as in the corresponding electric dipole case. Thus, if a bar magnet is placed in a magnetic field of strength H, with its axis normal to H, as shown in Fig.3.10(b), the two equal and opposite parallel forces $F=m.H$ acting on the two poles of the magnet form a couple, and the turning *moment* or torque of the couple, $T=F.l=ml.H=p_m.H$. If $H=1$, then $T=p_m$. The concept of magnetic dipole moment is of particular importance in the atomic study of magnetic materials.

3.10 Magnetic Shell: If the sectional area of a bar magnet becomes much greater compared with its length, a *magnetic shell* is formed. In other words, a magnetic shell is a thin sheet of iron magnetised along its thickness so that the poles appear on its two opposite surfaces. The idea of magnetic shell is often employed in developing the equivalence between a magnet and an electromagnet, Sec.4.8. Fig.3.11(a) shows a magnetic shell of thickness l, with N and S poles of strength m distributed uniformly over its area A. Fig.3.11(b) shows the pattern or profile of the associated lines of force.

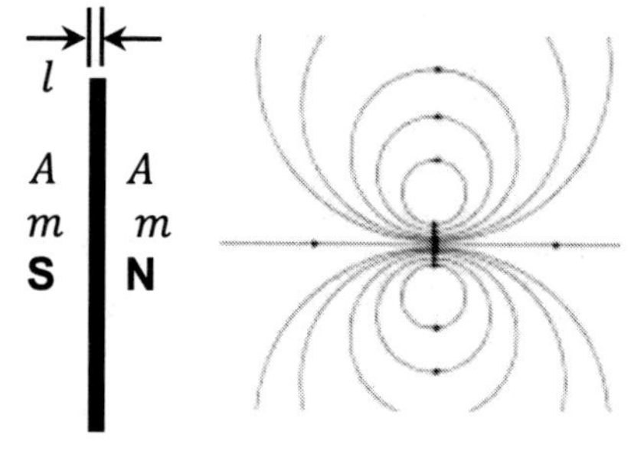

(a) (b)

Fig.3.11 Magnetic shell

A magnetic shell is specified by its *strength* which is defined as its magnetic moment per unit area, given by

$$\frac{p_m}{A} = \frac{ml}{A} = J.l \qquad (3.21)$$

where $J = \frac{m}{A}$ is its intensity of magnetisation.

3.11 Magnetisation of the Iron: Magnetisation of iron (magnetic material) is analogous to electric polarization of dielectric. An external magnetic field induces huge amount of magnetism in the entire volume of a specimen of iron along the direction of the field. The phenomenon is referred to as '*Magnetisation* or *Magnetic Polarization*' of iron. As such, the iron becomes a magnet, or is said to be '*magnetised*'. Consequently, some equal and opposite induced magnetic poles appear on the two opposite surfaces of the iron. Atomic interpretation of this phenomenon is not essential at this stage of study, so we skip the same.

Referring to Fig.3.12, let l be the length of the iron along the direction of the external magnetic field H, and A be its uniform cross sectional area. Under the influence of the external field, some equal N and S poles of strength m_i appear on the two opposite ends of the iron, as shown. Magnetisation may thus be conceived as induced magnetism in iron. Assuming that the entire volume of the iron is uniformly magnetised, the surface density of such magnetised pole per unit cross sectional area of the iron was defined in Sec.3.6 as the intensity of magnetisation J. We have

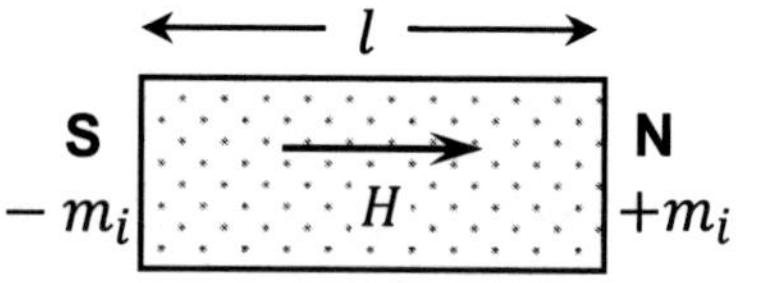

Fig. 3.12
Magnetisation of iron

$$J = \frac{m_i}{A} = \frac{m_i\, l}{Al} = \frac{p_m}{V} \qquad (3.22)$$

where $p_m = m_i l$ is the magnetic moment of the volume $V = Al$ of the iron. Thus, the intensity of magnetisation of a magnet may also be defined as its magnetic moment per unit volume.

To distinguish the related terms, (a) magnetic moment gives the strength of a magnet (b) magnetic moment per unit area gives the strength of a magnetic shell, and (c) magnetic moment per unit volume gives the intensity of magnetisation of a magnet. Owing to the hysteresis property of the iron, Ch.6, the induced magnetism does not fully disappear when the external field is withdrawn, but a '*residual magnetism*' remains in the iron. Hence, magnetisation is a *partially* reversible process, and so differs from electric polarization in this aspect. We now consider the effect or consequence of magnetisation, again using two concepts, as was mentioned is Sec.1.11.

(A) m – concept:

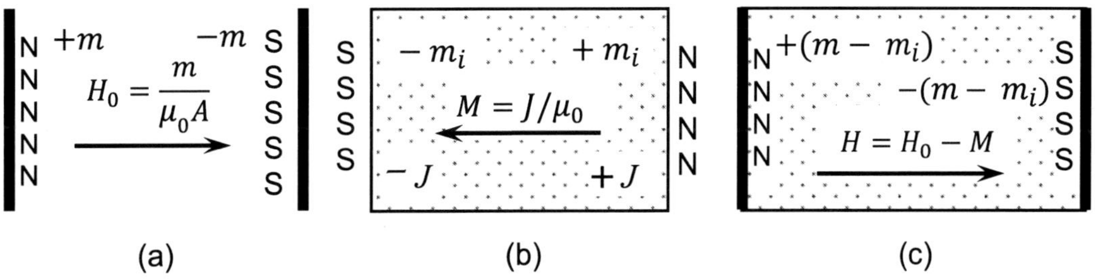

Fig.3.13 Magnetisation: m - concept

The m-concept is analogous to the Q-concept in Electrostatics. In Fig.3.13(a), the pole strengths $\pm m$ on the two close parallel N and S surfaces are uniformly distributed over the area A. With air gap between the surfaces, the field strength at any point between the surfaces, as was derived in Sec.3.7.1, $H_0 = m/(\mu_0 A)$. Let us then fill the air gap with some iron. Under the influence of the magnetic field, the iron is magnetised, Fig.3.13(b), and the polarities of the induced poles $\pm m_i$ being opposite to those of the main poles $\pm m$ on the surfaces, the net poles become $\pm(m - m_i)$, Fig.3.13(c). The field strength $M = \frac{m_i}{\mu_0 A} = \frac{J}{\mu_0}$ is defined as a vector quantity '*Magnetisation*'. As the field M acts in opposition to the main field H_0 , the net field in the iron reduces to $H = H_0 - M$. However, the amount of flux between the surfaces $\phi\ (= m)$ and the flux density $B\ (= \phi/A = m/A)$ do not change due to the insertion of the iron. Same as in Electrostatics, this implies that the induced pole is not associated with any flux, which is in compliance to the arbitrary definition of the flux. Again, the flux density at any point in the field, $B = \mu_0 H_0 = \mu H$.

Hence,

$$B = \mu_0 H_0 = \mu_0(H + M) = \mu_0 H + \mu_0 M$$

$$= \mu_0 H + J \qquad \text{.......} \quad (3.23)$$

This equation is analogous to Eqn.1.37 in Electrostatics. The quantity J in Eqn.3.23 is a flux density vector, whereas it is a pole density scalar in Eqn.3.22, as was indicated earlier in Sec.3.6. Recognition of J as flux density vector is more common than as pole density scalar. The quantity J as flux density is also referred to as '*ferric induction*' or '*intrinsic flux density B_i*'. The total flux density B is called as '*magnetic induction'*. These quantities should be recognised by these names.

(B) H – concept:

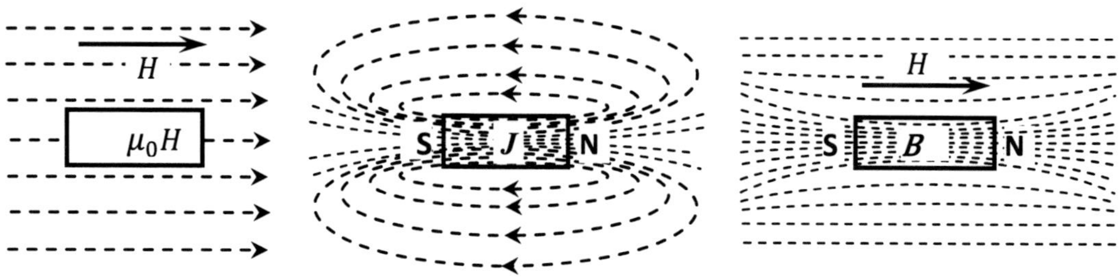

Fig.3.14 Magnetisation: H - concept

It is felt appropriate to consider this electromagnetic concept at this place. Whereas the magnetic field, in Magnetism, is conceived as produced by imaginary magnetic poles which have no physical existence, the same, in Electromagnetism, is actually produced by some electric current. Furthermore, whereas the field strength due to magnetic poles depends on the medium (its permeability), that due to electric current remains the same in all media. In fact, this is the fundamental distinction between Magnetism and Electromagnetism in the production of magnetic field. The properties of the magnetic field, however, remain the same, no matter how it is produced. Figs.3.14 and 3.15 describe the H- concept which is analogous to the E- concept in Electrostatics. In Fig.3.14(a), the magnetic field H is produced by some electric current through, say a solenoid (not shown in the figure), and the space is air. The flux density in the rectangular air bar shown, $B_0 = \mu_0 H$. If the bar is of iron, it becomes a bar magnet with its own lines of force shown in Fig.3.14(b). The lines of force of this magnet can be seen to act in conjunction with the main field within the iron, but in opposition outside it. This results in diverting most of the lines of force from outside the bar to inside it, as shown in Fig.3.14(c), which means that the magnetic flux density inside the bar is strengthened by many times. The self flux density within the magnet, Fig.(b), is given by its intensity of magnetisation J, and so the total flux density, Fig.(c), $B = B_0 + J = \mu_0 H + J$, which is Eqn.3.23.

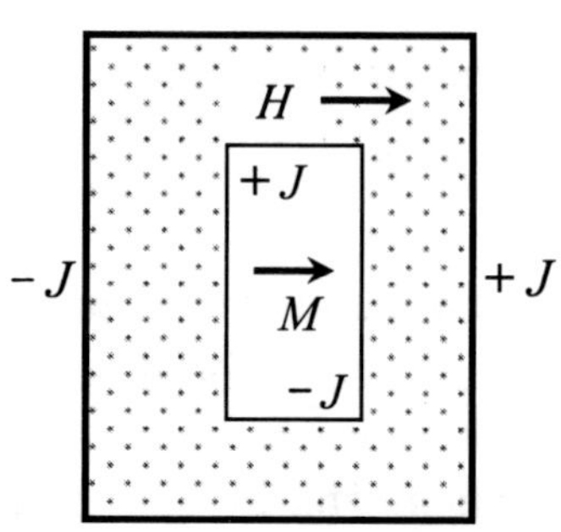

Fig. 3.15 Magnetisation: H- concept

Alternatively, if we consider a narrow imaginary cavity (air column) within the iron, as shown in Fig.3.15, the magnetisation field strength within the cavity $M = J/\mu_0$ acts in the same direction as that of H. Hence, the *total or effective* field strength becomes

$$H_E = H + M$$

$$= H + \frac{J}{\mu_0} = \frac{\mu_0 H + J}{\mu_0}$$

and the total flux density, $B = \mu_0 H_E = \mu_0 H + J$

In similarity with Electrostatics, Eqn.3.23 implies that the total flux density B (magnetic induction) in an iron may be thought of as comprising of two components – the flux density $\mu_0 H$ due to the external field H, plus the induced flux density J (ferric induction).

3.12 Magnetic Susceptibility and Magnetisation Curves of Iron:

3.12.1 Magnetic Susceptibility: Magnetic susceptibility is analogous to electric susceptibility, and hence the present discussion follows similar lines. In a magnetic material, ignoring for the moment its non-linear magnetic characteristic discussed below, the intensity of magnetisation J, hence the magnetisation $M(=J/\mu_0)$ and the ferric induction J, all can be taken to vary directly or linearly with the external magnetic field strength H. The proportionality factor defines a property of the magnetic material called its '*magnetic susceptibility*' denoted *by* χ_m, or simply by χ only, as the suffix m is often omitted. Hence, the definition of magnetic susceptibility depends on the quantity chosen. The quantity magnetisation field strength M is commonly chosen, and as such, magnetic susceptibility is defined as the unit-less ratio

$$\chi_m = \frac{Magnetisation\ field\ strength,\ M}{Magnetic\ field\ strength\ producing\ the\ magnetisation,\ H} = \frac{J/\mu_0}{H}$$

$$= \frac{J}{\mu_0 H} \qquad \text{.. (3.24)}$$

Magnetic susceptibility is related to permeability. Substituting for J from Eqn.3.24 in Eqn.3.23, we get $B = \mu_0 H + J = \mu_0 H + \mu_0 H.\chi_m = \mu_0 H(1 + \chi_m)$.

We also have $B = \mu_0 \mu_r H$.

Comparing these two relations, we get

$$\mu_r = \chi_m + 1 \quad \text{or, } \chi_m = \mu_r - 1 \qquad \text{........................... (3.25)}$$

Eqn.3.25 is analogous to Eqn.1.39 in Electrostatics. Defined as above, χ_m is unit-less number.

Sometimes magnetic susceptibility is defined as the ratio

$$\chi_m = \frac{Ferric\ induction,\ J}{Magnetic\ field\ strength\ producing\ the\ ferric\ induction,\ H} = \frac{J}{H}$$

When so defined, magnetic susceptibility is related with permeability and relative permeability by the relation

$$\chi_m = \mu - \mu_0 = \mu_0(\mu_r - 1)$$

and it bears the same unit of permeability, H/m.

Now, as was mentioned in Sec.3.2, μ_r of magnetic materials varies in hundreds & thousands. As such, the former definition of χ_m, Eqn.3.25, implies that the values of μ_r and χ_m are practically the same for iron. With the latter definition of χ_m, also, μ and χ_m are practically the same. This point leads to an important fact. By Eqn.3.24, $J = \mu_0 \chi_m H \approx \mu_0 \mu_r H$. Again, $B = \mu_0 \mu_r H$. This in turn implies that in Eqn.3.23, $B = \mu_0 H + J$, the term $\mu_0 H$ is practically negligible compared to J or B, and so $B = J$. This is really the case in practice with magnetic materials, as substantiated below with practical data.

3.12.2 Magnetisation curves of iron: Curves showing graphically how the various quantities J, B, μ_r and H vary with one another, generally with H plotted along abscissa, are collectively referred to as *magnetisation curves*. Hence, a magnetisation curve may be a J-H curve, a B-H curve, a μ_r - H curve, etc. The magnetisation curves shown in Fig.3.16(a) are not to scale, nor they represent any particular class of iron; these curves have been drawn only for illustration of the subject.

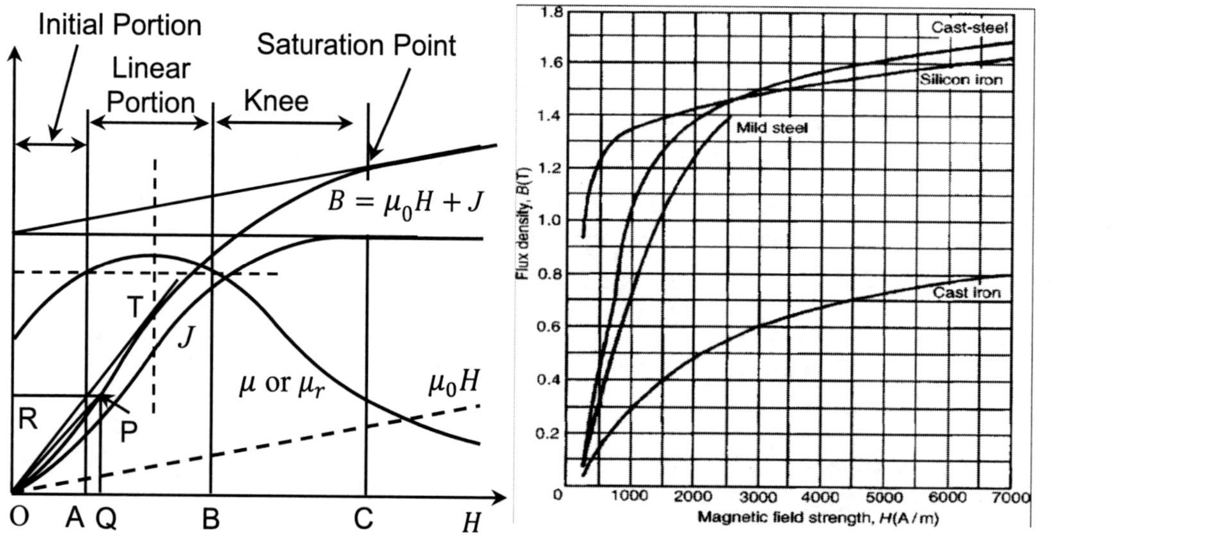

(a) Illustrative (b) Practical

Fig. 3.16 Magnetisation curves of magnetic materials

The (H versus) $\mu_r H$ curve is an inclined straight line and requires no elaboration. The profiles of the rest curves are outlined below. When a piece of some virgin iron (never magnetised before) is magnetised for the first time, for lower values of H from zero to A (initial portion of the curves), both J and B increase at a higher rate than H. The J and B curves are thus concave upwards. As $\mu = \mu_0 \mu_r$ and μ_0 is constant, the μ and μ_r curves are the same, only differing in scale by μ_0 times. It is, however, usual and convenient to consider the μ_r curve only. At $H = 0$, μ_r is not zero but has some appreciable value. At any point like P on the B-H curve, μ $(= B/H)$ is simply given by the slope of the line OP. Hence, μ_r increases in the initial portion. For values of H between A and B (linear portion), J and B curves are almost linear and very step, so μ_r remains almost constant at its maximum value. As H increases above the value B, magnetic saturation of the iron starts. The rate of increase of both J and B decreases sharply. This makes both J and B curves convex upwards, and μ_r falls off rapidly. Finally, at $H =$ C (saturation point), magnetic saturation completes, and J attains its highest value. Portion of the curves for $H =$ B to C is referred to as the 'knee' of the magnetisation curve. In fact, the J and B curves are concave upwards upto some point near the middle of the linear portion, beyond which they are convex upwards. At this transition point T, the line OT becomes tangent to the B - curve. The slope of the permeability line OP is highest at this point, and this represents the maximum value of μ_r. For values of H above C, J remains constant, so the B curve increases linearly following $B = \mu_0 H + J$, and with the same slope as that of $\mu_0 H$. It was explained in the previous Sec.3.12.1 that $J \approx \mu_0 \mu_r H$. Hence, as J remains constant and μ_0 is also constant, μ_r falls inversely with rising values of H. The shapes of these magnetisation curves are better explained in terms of the 'current-ring theory of magnetism', which we skip here. The understandings developed in this section can be supplemented with some typical magnetisation curves shown at Fig.3.16(b). Let us study the B-H curve of cast steel. At $H = 500$ A/m, $B = 0.45$ T. Hence, $\mu_0 H = 1.257\text{x}10^{-6}$ x 500 T= 0.00063 T, and $\mu_r = B/(\mu_0 H) = 716$. These numerical values comprise the second row of the following table.

Point on B-H curve	Value of H (in A/m)	Value of B (in T)	$\mu_0 H$ (in T) (calculated)	$\mu_r = B/\mu_0 H$ (calculated)	Portion on B-H curve
1	300	0.2	0.000377	530	Initial
2	500	0.45	0.00063	716	Linear
3	800	0.95	0.001	944	Maximum
4	1000	1.05	0.00126	835	Linear
5	2500	1.45	0.00314	461	Knee
6	3000	1.5	0.00377	398	Saturation
7	4700	1.6	0.0059	271	Saturated

Six other points on the curve have also been considered, and the results included in the table. The last column of the table has been filled in based on the judgment how μ_r varies with H. The calculated values of μ_r are in fair agreement with the fact that μ_r of cast steel varies in the range of 300–900. It can be seen in the above table that $\mu_0 H$ is too insignificant compared with B, as was obtained theoretically in previous Sec.3.12.1. Hence, the J and B curves are practically the same curve that becomes almost horizontal from the point of saturation. In fact, the slope of the $\mu_0 H$ curve in Fig.3.16(a) has been shown many times exaggerated for better clarity. Another point is that the 'initial portion' of Fig.3.16(a) is not visible in Fig.3.16(b). This owes to the fact that in Fig.3.16(a), the scale of the H - axis in the 'initial portion' has been highly widened for better clarity. As μ_r varies widely with H and B, specifying some value of μ_r for any magnetic material remains incomplete unless stated against some value of H or B.

To achieve most effective magnetic utilization of the iron, magnetic circuits of majority of electrical equipment are designed to operate within the linear portion of the magnetization curve with maximum range of permeability, and *by permeability of a material we generally mean the average permeability within this linear portion*. However, advantage of magnetic saturation is also taken in some applications like the cores of current transformers for metering purpose, to protect the instruments against damage by high inrush or fault currents. It also follows that so long any magnetic material operates within the linear portion of its magnetization curve, its permeability and susceptibility may be regarded as constants.

3.13 Force Between Magnetised Surfaces: Force exists between magnets or magnetised surfaces. Let us consider the simplest case – the attractive force between two close parallel N and S surfaces, Fig.3.17. The derivation is exactly similar to that for the attractive force between two oppositely charged parallel plates discussed in Sec.1.8.5. It was derived in Sec.3.7.1 that the field strength at any point between the two surfaces of area A each, neglecting the 'edge effect', due to *any one surface*, say, the N-surface, is $m/(2\mu A)$.

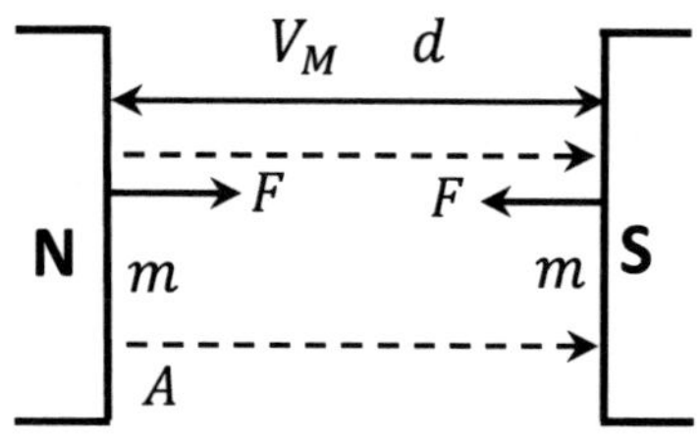

Fig. 3.17 Force between parallel N and S surfaces

This is also the field at all the points on the S-surface. This means, the entire S-surface, carrying a pole of total strength m, lies in a field $m/(2\mu A)$. Hence the force F exerted by the N-surface on the S-surface is $\{m/(2\mu A)\}.m = m^2/(2\mu A)$. This is also the force exerted by the S-surface on the N-surface. Also, since $m/A = J = B = \mu H$, the force of attraction between two close and parallel N and S surfaces, in various forms, is given by

$$F = \frac{m^2}{2\mu A} = \frac{J^2 A}{2\mu} = \frac{B^2 A}{2\mu} = \frac{\mu H^2 A}{2} = \frac{BHA}{2} \quad \text{............... (3.26)}$$

This equation is similar to Eqn.1.40 for the attractive force between two plates of a parallel plate capacitor. Corresponding to the capacitance C in Electrostatics, no analogous magnetic quantity is there. However, it will be seen in Electromagnetic Induction (Sec.5.4) that the inductance L is the analogous electromagnetic quantity.

3.14 Energy Stored in Magnetic Field: Energy stored in magnetic field SE_M is analogous to energy stored in electric field SE_E, Sec.1.14, and differs from the magnetic potential energy of an isolated pole in a magnetic field PE_M, Sec.3.8. *The amount of work done or energy expended by some external agency in creating a magnetic field is stored as magnetic potential energy SE_M in the field.* With isolated pole concept, the two oppositely magnetised parallel surfaces of Fig.3.17 may be thought of as created by gradual transference of $+m$ amount of magnetic pole from the S-surface to the N-surface across the air gap. As such, following the same line of discussion as for the energy stored in electric field, an identical set of equations can be developed for SE_M under alternative concepts and considerations. Thus, Fig.3.18 corresponds to Fig.1.36 in Electrostatics.

(A) Consideration – 1 (m – concept): At some instant t during the pole transference process from the S-surface to the N-surface in Fig.3.17, the instantaneous pole strengths $\pm m_t$ on the surfaces create, by Eqn.3.17, an instantaneous magnetic PD $v_M = m_t d/(\mu A)$ between the surfaces. v_M thus increases linearly with m_t, as shown at Fig.3.18(a). Work done or increase in SE_M in transferring dm_t amount of N-pole pole from the S-surface to the N-surface in time dt is given by

$$d(SE_M) = v_M.\ dm_t \qquad \text{.................................. (3.27)}$$

represented by the area of the elemental strip shown. The SE_M in the magnetic field created by the poles $\pm m$ and to magnetic PD V_M is, therefore,

$$SE_M = \int_{m_t=0}^{m_t=m} d(SE_M) = \int_0^m v_M.\ dm_t = \frac{d}{\mu A}\int_0^m m_t.dm_t = \frac{d.m^2}{2\mu A}$$

$$= V_M.m/2 \qquad \text{................................ (3.28)}$$

which is the area of the dotted triangle OAm *below* the V_M-m curve.

As was discussed in the corresponding electric case, Sec.1.14, SE_M is always positive, whereas PE_M can be positive or negative.

(B) Consideration – 2 (m – concept): During the said pole transference process, as the magnetic PD across the surfaces v_M increases from zero to V_M uniformly (linearly) with the amount of pole transferred m_t, the average PD is $V_M/2$. So, work done in transferring m amount of pole under this average PD, which is the SE_M in the field, is given by $SE_M = V_M m/2$.

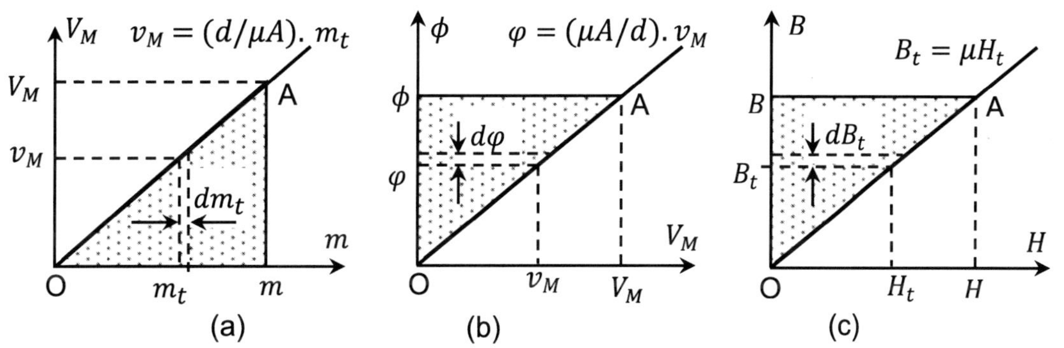

Fig. 3.18 Energy stored in magnetic field

(C) Consideration – 3 (ϕ – concept): Magnetic flux ϕ between the N and S surfaces being numerically equal to the pole strengths $\pm m$ on the surfaces, Eqn.3.27 may be written as

$$d(SE_M) = v_M.d\varphi \qquad \text{.................................. (3.29)}$$

with graphical representation as in Fig.3.18(b), wherein the instantaneous flux $\varphi\ (= m_t)$ increases linearly with v_M. As such,

$$SE_M = \int d(SE_M) = \int v_M.d\varphi = \frac{V_M.\phi}{2}$$

$$= \left[\frac{PD.flux}{2}\right] \qquad \text{.............................. (3.30)}$$

and is given by the area of the dotted triangle OAϕ *above* the ϕ-V_M curve.

These two alternative concepts sense differently. The former concept implies that the pole strength (cause, along X-axis) results in the magnetic PD (effect, along Y-axis). On the other hand, the latter concept implies that the magnetic PD (cause, along X-axis) results in the magnetic flux (effect, along Y-axis). The former concept implies that work has to be done, or energy has to be expended, in creating opposite poles on the two surfaces. The latter concept implies that creation of magnetic field involves expenditure of energy. Both these concepts are applicable in Magnetism, but only the latter concept in Electromagnetism.

Expanding the relation $d(SE_M) = v_M.d\varphi$ as

$$d(SE_M) = v_M.d\varphi \; = (H_t.d).(A.dB_t)$$
$$= (A.d).(\,H_t.dB_t)$$

where H_t and B_t are respectively the instantaneous magnetic field and magnetic flux density at any point between the surfaces, and $(A.d)$ is the volume of the field between the surfaces, the increase in energy stored per unit volume in the field during time dt

$$d(SE_m) = \; H_t.dB_t \qquad \text{................................ (3.31)}$$

as represented graphically in Fig.3.18(c). Here, $B_t\,(= \mu.H_t)$ increases linearly with H_t. Hence, energy stored per unit volume of the field with field strength H and flux density B

$$SE_m = \int d(SE_m) = \int H_t.dB_t = \frac{HB}{2}$$
$$= \left[\frac{field\ strength.flux\ density}{2}\right]$$

This is given by the area of the dotted triangle OAB *above* the B-H curve. This representation of energy stored per unit volume in terms of the area of the B-H curve helps deriving the hysteresis loss in magnetic materials in terms of the area of the hysteresis loop, Sec.6.1.4. Using the relation $B = \mu H$ that presents H as the cause and B as its effect, as was mentioned in Sec.3.5, the expression for the energy stored per unit volume in magnetic field in different forms is

$$SE_m = \frac{HB}{2} = \frac{\mu H^2}{2} = \frac{B^2}{2\mu} \qquad \text{................................ (3.32)}$$

This generic field equation holds true for any point in a magnetic field of any configuration. In case of parallel or uniform field, SE_m is the same at all points in the field. In non-uniform fields, it varies from point to point in the field.

Two more concepts of SE_M will be developed. These are ϕ-concept in Electromagnetism (Sec.4.7.1) and an alternative I- concept in Electromagnetic Induction (Sec.5.4).

Table 3.1: Analogy and Distinction between Electrostatics and Magnetism

<table>
<tr><th colspan="2">Electrostatics</th><th colspan="2">Magnetism</th></tr>
<tr><td colspan="4">Analogy</td></tr>
<tr><td>Quantity</td><td>Unit</td><td>Quantity</td><td>Unit</td></tr>
<tr><td>Electric Charge (Q)
$F = \frac{Q_1 Q_2}{4\pi\varepsilon x^2}$</td><td>coulomb (C)</td><td>Magnetic Pole (m)
$F = \frac{m_1 m_2}{4\pi\mu x^2}$</td><td>weber (Wb)</td></tr>
<tr><td>Permittivity (ε)</td><td>farad/metre (F/m)</td><td>Permeability (μ)</td><td>henry/metre (H/m)</td></tr>
<tr><td>Electric Field Strength (E) $E = F/Q$</td><td>volt/meter (V/m)</td><td>Magnetic Field Strength (H)
$H = F/m$</td><td>ampere/metre (A/m)</td></tr>
<tr><td>Electric Flux (Ψ)
$\Psi = Q$</td><td>coulomb (C)</td><td>Magnetic Flux (ϕ)
$\phi = m$</td><td>weber (Wb)</td></tr>
<tr><td>Electric Flux Density (D) $D = \psi/A = \varepsilon E$</td><td>coulomb / sq. meter (C/m^2)</td><td>Magnetic Flux Density (B) $B = \phi/A = \mu H$</td><td>weber / sq. metre (Wb/m^2) or tesla (T)</td></tr>
<tr><td>Electric Charge Density (σ) or Electric Polarization as charge density (P)
σ or $P = Q/A$</td><td>coulomb /sq. metre (C/m^2)</td><td>Intensity of Magnetisation (J)
$J = m/A$</td><td>weber /sq. metre (Wb/m^2)</td></tr>
<tr><td>Electric Potential or PD (V)
$V = -\int E.dx$</td><td>volt (V)</td><td>Magnetic Potential or PD (V_M)
$V_M = -\int H.dx$</td><td>ampere (A)</td></tr>
<tr><td>Electric Dipole Moment
(p_e) $p_e = Qd$</td><td>coulomb metre (C.m)</td><td>Magnetic Moment
(p_m or p) $p_m = ml$</td><td>weber metre (Wb.m)</td></tr>
<tr><td>Electric Polarization as flux density (P)
$P = \varepsilon_0 \chi_e E$</td><td>coulomb /sq. metre (C/m^2)</td><td>Ferric induction (J)
$J = \mu_0 \chi_m H$</td><td>weber /sq. metre (Wb/m^2) or tesla (T)</td></tr>
<tr><td>Electric Susceptibility (χ_e) $\chi_e = P/(\varepsilon_0 E)$</td><td>Unit-less</td><td>Magnetic Susceptibility (χ_m or χ)
$\chi_m = J/(\mu_0 H)$</td><td>Unit-less</td></tr>
</table>

<table>
<tr><td colspan="2">Electrostatics</td><td colspan="2">Magnetism</td></tr>
<tr><td colspan="4">Analogy</td></tr>
<tr><td>Quantity</td><td>Unit</td><td>Quantity</td><td>Unit</td></tr>
<tr><td>PE of an electric charge (PE_E)
$PE_E = QV$</td><td rowspan="2">joule (J)</td><td>PE of a magnetic pole (PE_M) $PE_M = mV_M$</td><td rowspan="2">Joule (J)</td></tr>
<tr><td>Energy stored in electric field (SE_E)
$SE_E = QV/2$</td><td>Energy stored in magnetic field (SE_M)
$SE_M = mV_M/2$</td></tr>
<tr><td colspan="4">Distinction</td></tr>
<tr><td colspan="2">Electric charge can exist in isolation</td><td colspan="2">Magnetic pole cannot exist in isolation</td></tr>
<tr><td colspan="2">Electric flux lines may not form closed loops</td><td colspan="2">Magnetic flux lines always form closed loops</td></tr>
</table>

Ch. 4: Electromagnetism

4.1 Introduction: Electromagnetism deals with the magnetic effect of electric current. Electricity produces magnetism. In other words, every electric current is associated with a magnetic field. This property of electric current is known as its magnetic effect. While preparing for an evening lecture on heating effect of electric current on 21st. April 1820, Danish physicist Hans Christian Oersted (1777-1851) noticed that a magnetic compass needle lying near a conductor deflected whenever an electric current was passed through the conductor. As it was known by then that some magnet can only deflect a magnetic needle, it was logical to conclude that an electric current behaves like a magnet. Oersted's discovery laid the foundation of Electromagnetism. Andre Marie Ampere (1775-1836), a French physicist, mathematician and Fellow of the Royal Society, heard of Oersted's discovery on 11th. September 1820. Only a week later, on 18th. September 1820, he presented a paper to the Academy containing a far more complete exposition related to Electromagnetism, including the force between parallel currents. Ampere is regarded as one of the discoverers of Electromagnetism, and also as the Newton of Electricity. Electromagnetism thus redefines Magnetism in terms of electric current. It is rightly said that electricity and magnetism are not distinct phenomena, but are only two distinct manifestations of the same phenomenon, just like two opposite faces of the same coin. This saying will be gradually realised with the progress of this present chapter as well as the next chapter on Electromagnetic Induction.

For convenience in distinguishing between the various topics, the contents of this chapter have been grouped under three parts.

Part A: Action of Electric Current on Magnet

4.2 Force of Electric Current on Magnetic Pole: Following Oersted's discovery, two French physicists, Jean-Baptiste Biot (1774-1862) and Felix Savart (1791-1841) investigated the subject further. In October 1820, they announced that the force exerted by a long and straight current carrying conductor on a magnetic pole, as measured by compass needle, varies inversely to the perpendicular distance of the pole from the conductor. A few days after this announcement, French mathematician, astronomer and physicist, Pierre-Simon de Laplace (1749-1827), who is regarded as the Newton of France, supplemented analytically the above empirical proportionality law in an accurate mathematical equation

$$df = \frac{m.I.dr.Sin\,\theta}{4\pi l^2} \qquad \text{.. (4.1)}$$

where, as shown in Fig.4.1, df is the elemental force exerted by an elemental length dr of a conductor carrying a current I on a north pole m situated at point P distant l from dr, and θ is the angle between the distance line l and the tangent to the elemental length dr. Eqn.4.1 is referred to as *Biot-Savart-Laplace Law*. As regards the direction of this force df, if the elemental length dr and the point P lie on the plane of the paper, as shown, the direction of the force on the north pole, for the shown direction of current, is perpendicular to the plane of the paper and *inward* it, as represented by the encircled cross ⊕. (An encircled dot ⊙ also represents a direction perpendicular to the plane of the paper but *outward* or *away from* it. The dot and cross represent respectively the head and tail of an arrow.)

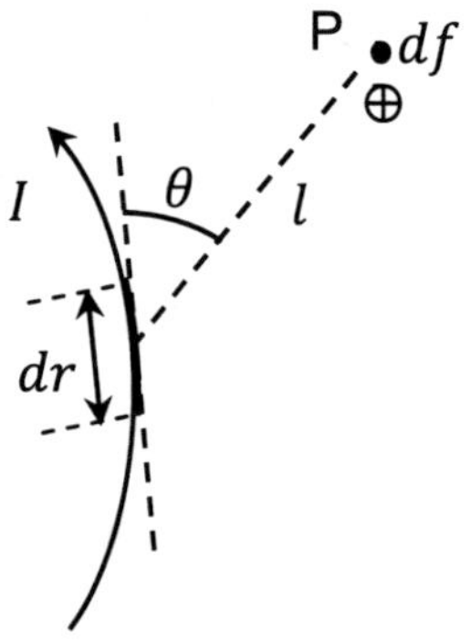

Fig.4.1 Biot-Savart-Laplace Law

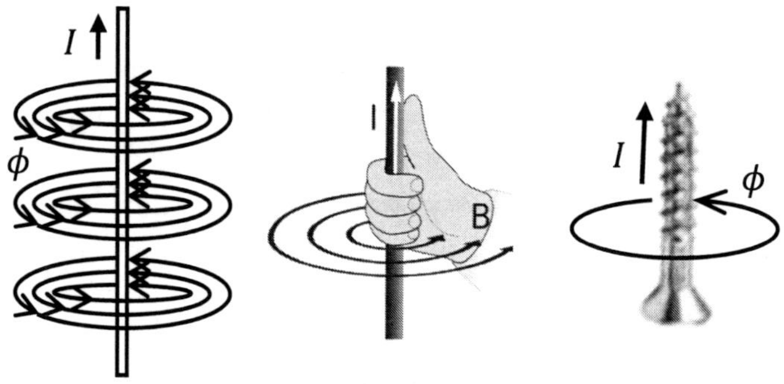

(a) (b) (c)

Fig.4.2 Magnetic field around straight current

It therefore follows that the lines of force surrounding a straight current carrying conductor, Fig.4.2(a), are innumerable number of coaxial circles about the axis of the conductor with all possible radii and lying on all possible planes perpendicular to the axis of the conductor. The direction of these lines of force is best remembered by *Right Hand Thumb Rule*, Fig.4.2(b) or *Right-Handed Screw Rule*, Fig.4.2(c).

Magnetic field strength being the force on a unit north pole or 'test' pole, it follows from Eqn. 4.1 that the magnetic field strength at point P in Fig.4.1

$$dH = \frac{I.dr.Sin\,\theta}{4\pi l^2} \qquad \text{...} (4.2)$$

One point should be noticed at this stage. Whereas the force between magnets (or magnetic poles) is dependent on the medium (permeability), that between electric current and magnet (or magnetic pole) is independent of the medium.

4.3 Magnetic Field near Straight Current: In Fig.4.3, field at point P due to the elemental length dr of the straight current I, as given by Eqn. 4.2,

$$dH = \frac{I.dr.Sin\,[(\pi/2)\pm\phi]}{4\pi l^2} = \frac{I.dr.Cos\phi}{4\pi l^2}$$

Now, $dr.Cos\,\phi = l.d\phi$ and $Cos\,\phi = x/l$,

or, $\frac{dr}{l} = \frac{d\phi}{Cos\,\emptyset} = \frac{d\phi}{x/l}$ or, $\frac{dr}{l^2} = \frac{d\phi}{x}$

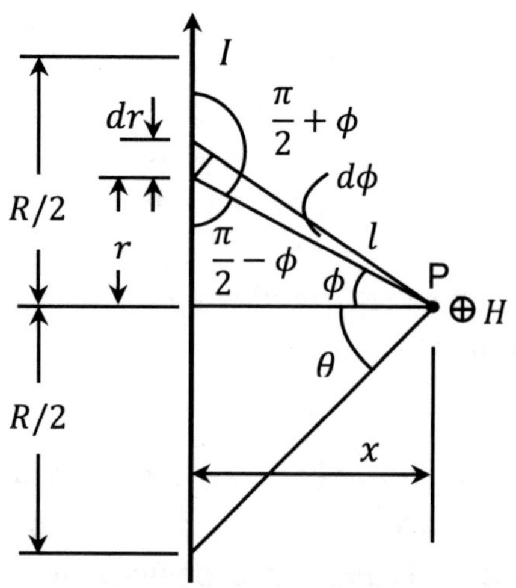

Fig. 4.3 Magnetic field strength near straight current

So, $dH = \frac{I}{4\pi x} Cos\,\phi\,.d\phi$ and the field at P due to the shown length R of the conductor,

$$H = 2\frac{I}{4\pi x}\int_{\phi=0}^{\phi=\theta} Cos\,\phi.\,d\phi = \frac{I}{2\pi x}\,[\,Sin\phi\,]_0^{\theta}$$

$$= \frac{I}{2\pi x} Sin\theta = \frac{I}{2\pi x}\left[\frac{R/2}{\sqrt{(R/2)^2+x^2}}\right] \quad \text{.................... (4.3(a))}$$

When the length of the conductor R is very great compared to the distance x, ($R \gg x$), $\theta = \pi/2$, so $Sin\,\theta$ (or the term within the square brackets) = 1, and then

$$H = \frac{I}{2\pi x} \quad \text{.................................... (4.3)}$$

Eqn.4.3 is analogous to Eqn.1.9 in Electrostatics. Eqn.4.3 shows that $H \propto 1/x$, as was found earlier experimentally by Biot and Savart. This field being the same along the circumferential length $l = 2\pi x$ of the circle of radius x round the conductor, *work done in moving a test pole once round the current* I, which *is defined as the magneto-motive force (mmf) of the current (source)* I, is $F = H.2\pi x = I$. The mmf is thus independent of the radius x. Simply stated, the mmf of a current I is given by $F = I$. In general, if H varies with x, the generalised form of mmf is expressed as

$$F = \oint H.dx = I \quad \text{... (4.4)}$$

Eqn.4.4 explains that the unit of the mmf is ampere (A).

Work done or PD between two points being independent of the route followed in moving from one point to another, mmf of a current remains the same for any geometry of the route (line of force), which need not necessarily be circular, followed by the test pole in moving round the current. This is an important understanding.

Further, if there be N number of very closely-packed thin insulated conductors each carrying the same current I in the same direction, the total or effective current becomes NI. As such, the field strength becomes $H = NI/2\pi x$ and the mmf becomes $F = NI$. The unit of mmf still remains ampere, as the number of turns N is a unit-less quantity.

4.4 Magnetic Field near Circular Current:

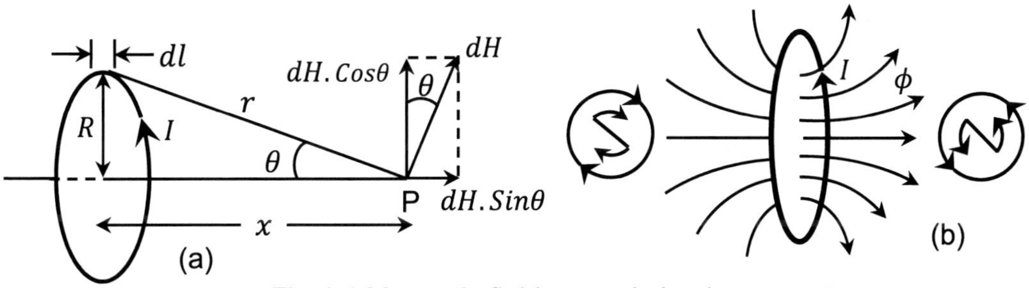

Fig.4.4 Magnetic field around circular current

A conductor bent to some closed or nearly-closed geometry is called a 'coil', and a coil carrying a current is referred to as a current loop. With Fig. 4.4(a), let us find the field at some point P on the axis of a circular current loop carrying current I. The plane of the coil is vertical, and normal to the plane of the paper. The situation being three-dimensional, the circle has been shown in its elliptical isometric view for clarity.

Elemental field strength at P due to the shown elemental length dl on the top of the coil, as given by Eqn.4.2, $dH = \frac{I.dl}{4\pi r^2}$. If this elemental field be split into its vertical and horizontal components, as shown, all such vertical components over the entire length of the coil cancel out, and the total horizontal field

$$H = \int_{l=0}^{l=2\pi R} \frac{I.dl.Sin\,\theta}{4\pi r^2} = \frac{I.2\pi R.Sin\,\theta}{4\pi r^2} = \frac{I.R.Sin\,\theta}{2r^2}$$

Since $Sin\,\theta = R/r$, we get for N number of turns on the coil

$$H = \frac{NIR^2}{2r^3}$$

$$= \left[\frac{NI}{2R}\right] Sin^3\,\theta \qquad \text{.................................... (4.5)}$$

For a given coil carrying some steady current, N, I and R are constants. As x decreases, r decreases and $Sin\,\theta$ increases, so H increases. At the centre of the circle, $r = R$, or $\theta = \pi/2$, so

$$H = \frac{NI}{2R} \qquad \text{.............................. (4.6)}$$

Profile of the magnetic lines of force around a circular current loop is shown at Fig. 4.4(b), which reveals that a current loop resembles a magnetic shell described in Sec.3.10. Magnetic polarity of a current loop, with reference to the direction of the current, is best remembered by the '*Clock Rule*' shown in Fig. 4.4(b) as encircled and arrowed 'N' and 'S'.

Table 4.1: Geometries of usual Coils and Windings

Coil Length	Turn	Conductor Section	Coil Section	Core Axis	Core Section
Short-length coils	Single	Circular or rectangular	-	-	Circular or rectangular
	Multi	Circular or rectangular	Circular or rectangular	-	Circular or rectangular
Long coils (Windings)	Multi	Circular or rectangular	Rectangular	Straight (Solenoid)	Circular or rectangular
				Circular	Circular or rectangular (Toroid)
					Circular (Torus)
Multi-turn coils wound on polygonal cores				Rectangular	Rectangular

Chart 4.1: Family of Commonly Used Shapes of Coils and Windings

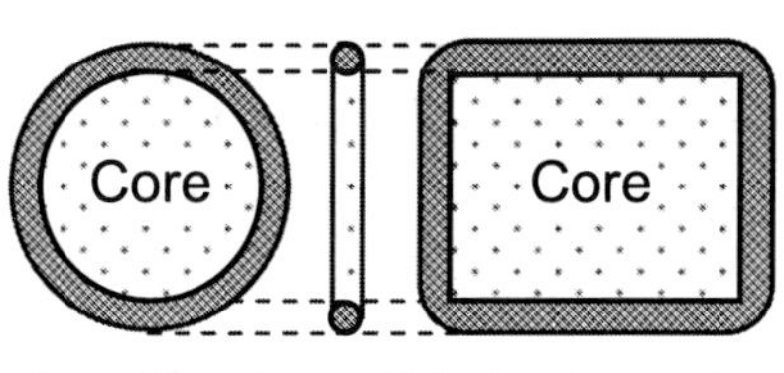

(a) Circular coil (b) Rectangular coil

Single-turn coils of circular conductor section

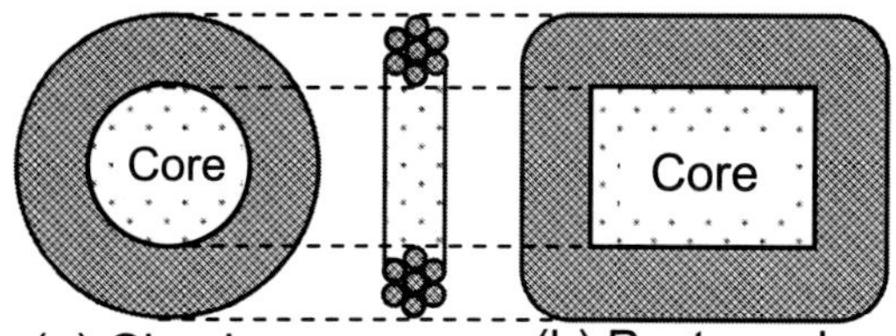

(a) Circular core section (b) Rectangular core section

Multi-turn (2-layer) coils of circular conductor and coil sections

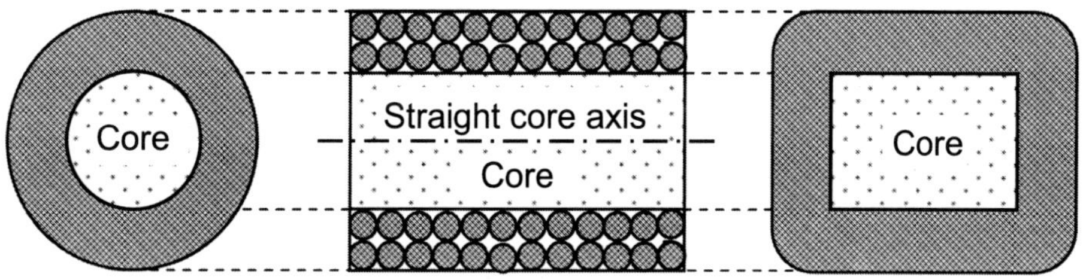

(a) Circular core section (b) Rectangular core section

Solenoids: Windings (2-layer) of rectangular coil section and straight core axis

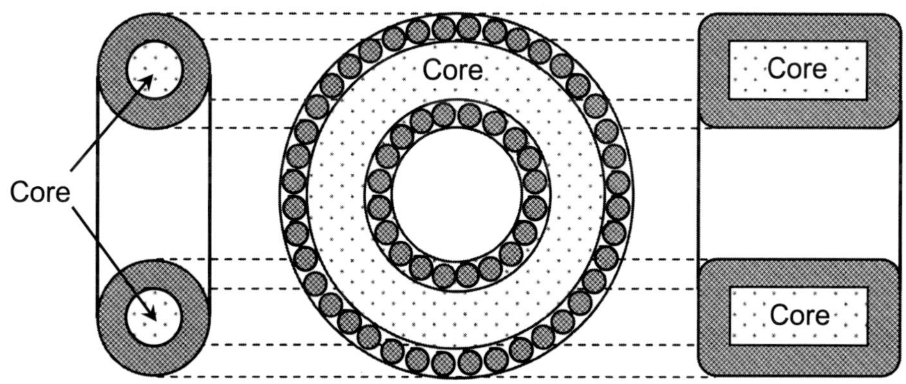

(a) Circular core section (Torus) (b) Rectangular core section

Toroids: Windings (1-layer) of rectangular coil section and circular core axis

(A "toroid" may be defined as the surface of a solid obtained by rotating an area about an axis on the plane of the area but outside it. If the area is circle, the solid becomes a "torus")

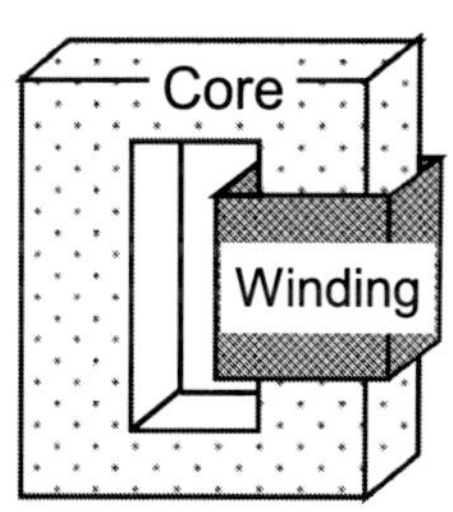

Coils wound on polygonal cores: These are general types of coils wound on cores with non-circular or polygonal axes, the shown core of rectangular axis and rectangular section being most common. Toroids are thus a special case of this general type, wherein the core axis is circular.

Field strength at points not lying on the axis of the circular current loop (Fig.4.4) will certainly be too complex to derive analytically, even approximately, if not impossible. Further, if the shape of the loop is not circular, the analytical derivation of the field strength is difficult or impossible even for points on the axis. Such complicated situations, however, do not much bother us. We consider only some simple cases, and generalise the results to cover the complicated cases by *logical reasoning*, with some acceptable approximations in some cases. These comments are perhaps equally applicable to all branches of science and technology.

Multi-turn coils wound over some core are called '*windings*'. Conductors are usually stored coiled on 'bobbins', subsequently wound or formed on 'formas', and then slid on cores. It is necessary to become familiar with the coils and windings of usual geometries. These have been explained by Table 4.1 and Chart 4.1.

4.5 Magnetic Field of Solenoidal Current:

Identical flux distribution patterns shown at Fig.4.5 reveal that a solenoidal current resembles a bar magnet. Fig.4.6(a) shows a solenoid, of axial length l, of circular core section of radius R, and wound closely and uniformly with N number of turns. Let I be the current through the turns. The number of turns per unit length, $n = N/l$. Let us derive the field strength H at the point P distant x from the centre of the solenoid and on its axis.

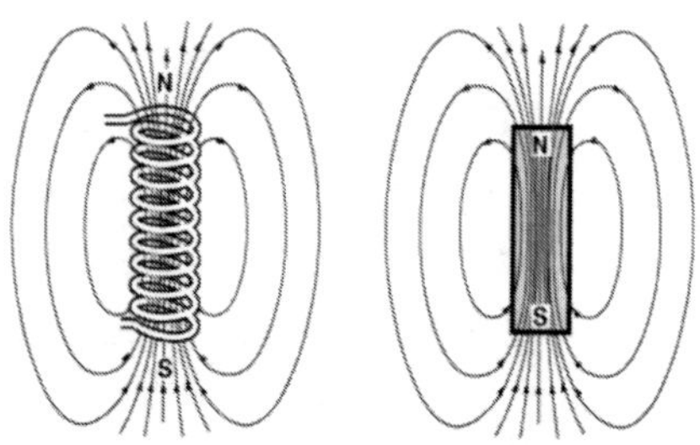

Fig.4.5 A solenoidal current resembles a bar magnet

The current in the elemental length dy of the solenoid being $n.I.dy$, the elemental horizontal field at P due to the elemental length dy of the solenoid, as given by Eqn.4.5,

$$dH = \frac{(n.I.dy)R^2}{2r^3} = \frac{n.I}{2}.\frac{R}{r}.\frac{dy.R}{r^2} = \frac{n.I}{2} Sin\,\theta.d\theta$$

(Length BC $= dy\,Sin\,\theta = r\,d\theta$ and $Sin\,\theta = R/r$. Hence, $dy\frac{R}{r} = r\,d\theta$, or, $\frac{dy.R}{r^2} = d\theta$)

The total horizontal field at P due to the entire solenoid, therefore,

$$H = \frac{n.I}{2}\int_{\theta_1}^{\theta_2} Sin\,\theta\,d\theta, \text{ where } tan\theta_1 = \frac{R}{x+(l/2)} \text{ and } tan\theta_2 = \frac{R}{x-(l/2)}$$

$$= \frac{n.I}{2}\left[-Cos\,\theta\right]_{\theta_1}^{\theta_2} = \frac{n.I}{2}(Cos\,\theta_1 - Cos\,\theta_2) = \frac{NI}{2l}(Cos\,\theta_1 - Cos\,\theta_2)$$

$$H = \frac{NI}{2l}\left[\frac{x+(l/2)}{\sqrt{[x+(l/2)]^2+R^2}} - \frac{x-(l/2)}{\sqrt{[x-(l/2)]^2+R^2}}\right] \quad \text{............. (4.7)}$$

Field at mid length of the solenoid

$$H_m = \frac{NI}{2l}2Cos\,\theta_3 = \frac{NI}{l}Cos\,\theta_3 = \frac{NI}{l}\frac{l/2}{\sqrt{(l/2)^2+R^2}} = \frac{NI}{2\sqrt{(l/2)^2+R^2}}$$

And the field at either end of the solenoid

$$H_e = \frac{NI}{2l}Cos\,\theta_4 = \frac{NI}{2l}\frac{l}{\sqrt{l^2+R^2}} = \frac{NI}{2\sqrt{l^2+R^2}}$$

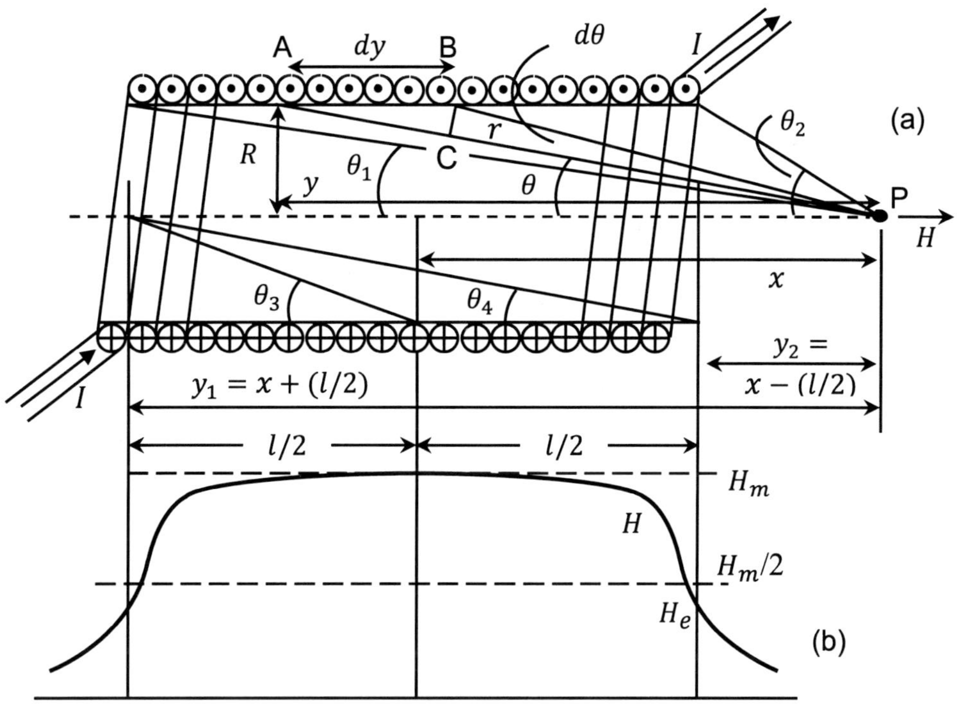

Fig.4.6 Magnetic field of solenoid

Clearly, $H_m > H_e$. When $l \gg R$, $H_m = NI/l$ and $H_e = NI/(2l)$. The trend of variation of the field strength H over the axis of the solenoid has been approximately shown at Fig.4.6(b). If, for example, the length is 10 times the diameter ($l = 20R$), the variation of H over 10% of the length on either side of the centre is only within about 1%. This means, with large l/R ratio, average field strength on the axis of a solenoid over most of the central portion, may, with reasonable accuracy, be written as

$$H_m = \frac{NI}{l} \qquad \text{............} \qquad (4.8)$$

and that over some limited end portions as

$$H_e = \frac{NI}{2l} \qquad \text{............} \qquad (4.9)$$

It can be noticed that the field strength is independent of the radius of the solenoid.

4.6 Magnetic Field of Toroidal Current: If the straight axis of the circular-section solenoid of Fig.4.6 be bent as a closed circle, the solenoid is converted into the torus shown at Fig.4.7. The axis of the torus being closed, any point on the axis does not see any end. As such, the torus behaves as a solenoid with infinite l/R ratio. Hence, as was discussed in Sec.4.5, the magnetic field strength at any point on the axis of a torus

$$H = \frac{NI}{l} \qquad \text{............} \qquad (4.10)$$

where, N is the number of closely and uniformly wound turns spread over the entire axial length l of the torus, and I is the current in the turns.
Hence, the mmf of the torus,

$$F = H.l = NI \qquad \text{.. (4.11)}$$

The same result was also obtained for straight current in Sec.4.3. Hence, by *logical reasoning*, we conclude that the mmf of N turns of current I is always NI, irrespective of the geometry of the coil or winding carrying the current. In the cases of a straight conductor and the torus the magnetic lines of force (path followed by a test pole) being of definite (circular) geometry, and also that the value of H remaining the same for all points on any line of force, we could easily derive the mmf. In other cases like the circular and solenoidal currents, Sections 4.4 and 4.5 respectively, where these two conditions do not prevail, the direct derivation of the mmf becomes difficult or impossible. In all cases, however, the generalised result $F = NI$ remains valid without any approximation.

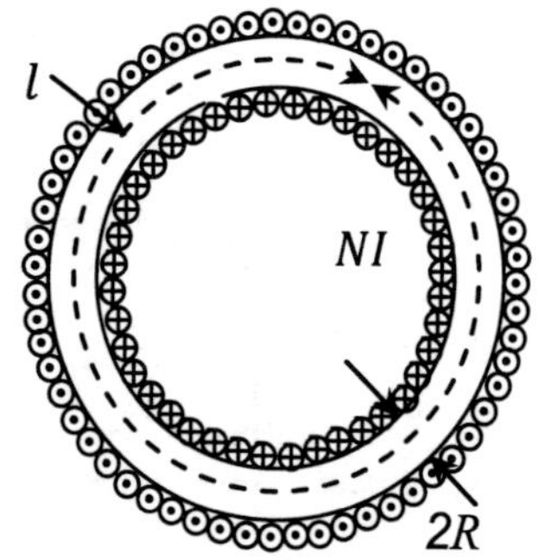

Fig.4.7 Magnetic field inside a torus

4.7 The Magnetic Circuit: The term 'circuit' means some closed path, and an electric circuit means a closed path for an electric current. Similarly, *a magnetic circuit may be defined as the closed path for a magnetic flux.* In the magnetic circuit of the torus of Fig.4.7, the field strength at any point on the axis of a torus $H = NI/l$, and the flux density $B = \mu H = \mu NI/l$. *Assuming that the average field strength, and so the average flux density, at all points on the entire sectional area of the torus remain the same as that on the axis*, the total flux through the torus
section becomes $\phi = B.A = \mu NIA/l$, where $A = 4\pi R^2$ is the sectional area of the torus.
Now $F = NI$ being the mmf, ϕ may be written as

$$\phi = \frac{NI}{l/(\mu A)} = \frac{F}{S} = F.P \qquad \text{.................................... (4.12)}$$

The quantity $S = l/(\mu A)$, dimensionally $1/(\mu l)$, is defined as the *reluctance* of the magnetic circuit, and its reciprocal $P = 1/S = \mu A/l$ as its *permeance*. Eqn.4.12 is analogous to the equation

$$I = \frac{E}{R} = \frac{E}{\frac{\rho l}{A}} = \frac{E}{\frac{l}{\gamma A}} = E\frac{\gamma A}{l} = EG$$

in Current Electricity, where R and G are respectively the resistance and conductance. It will be seen later (Sec.5.4) that dimensionally inductance is given by (permeability x length), and the unit of inductance being henry, reluctance is expressed in the unit of 'per henry' (H^{-1}). The relation

$$F = (NI = H.l =)\ \phi.S \qquad \text{... (4.13)}$$

is thus analogous to the relation $E = I.R$ in Current electricity.

Magnetic circuit of the torus is its core within which the magnetic flux exists. If this core is of iron instead of air, the amount of flux increases by μ_r times. This is analogous to the fact that electric current increases by using high-conductivity material in the electric circuit. Reluctances in series and parallel obey the same laws as those of resistances. Further, the electric circuit laws like Kirchhoff's current and voltage laws for composite (series-parallel) electric circuits also apply to composite magnetic circuits as corresponding flux and mmf laws.

The magnetic circuit is thus analogous to the electric circuit. There are, however, two aspects in which magnetic circuits differ from electric circuits:

(1) The laws of magnetic circuit are approximate; approximation creeps in by using some average value of the field strength and flux density over the core sectional area. Non-uniform magnetic property at various points within a given specimen of any class of iron also contributes to the approximation. Conversely, the laws of electric circuits are accurate. In fact, the term 'estimate' is perhaps more appropriate than the term 'formula' for Eqn.4.12.

(2) Energy is required to be expended in developing magnetic flux in magnetic circuit, and this energy is stored in the magnetic field. Once the magnetic flux is developed, no energy is required to be expended in maintaining the flux; the energy supplied to the magnetising coil is dissipated only as ohmic heating in the resistance of the coil. Conversely, in electric circuits energy is required to be supplied continuously in maintaining the current.

Electric and magnetic circuits have been compared in Table 4.2 on P- 4.14.

4.7.1 Energy Stored in Electromagnetic Field: By the term 'electromagnetic field', we mean the magnetic field produced by electric current, to distinguish from the 'magnetic field' produced by isolated poles. It was discussed in Sec.3.14(C) that within a volume of a *magnetic field*, if the flux through its area increases by $d\varphi$ at an instantaneous magnetic potential difference v_M across its length, the energy stored within the volume increases by $d(SE_M) = v_M.d\varphi$ (Eqn.3.29). In an *electromagnetic field*, say within the volume of the torus of Fig.4.7, an instantaneous current i creates an instantaneous mmf or magnetic PD $v_M = Ni$ round the axial length of the torus. Hence, the energy stored in the electromagnetic field increases by

$$d(SE_M) = Ni.d\varphi \quad \text{...} \quad (4.14)$$

If a steady current I creates a steady flux ϕ, steady mmf $V_M = NI$, and in line with Eqn.3.30,

$$SE_M = \frac{1}{2} NI\phi \quad \text{...} \quad (4.15)$$

This is the ϕ- concept in Electromagnetism, as was mentioned in Sec.1.14.

4.7.2 Flux law of magnetic circuit: Flux law in magnetic circuits is exactly analogous to Kirchhoff's current law in electric circuits. Flux law states that the algebraic sum of the magnetic fluxes *towards* or *away from* any junction region in a magnetic circuit is zero.

Mathematically this law is stated as $\sum\phi = 0$. Flux towards a junction is conventionally taken as positive and that away from the junction as negative. As such, this law may also be stated as that the total flux towards a junction equals the total flux away from the junction, stated mathematically as $\sum\phi_i = \sum\phi_o$, where the suffices i denotes incoming and o outgoing. Proof of this law simply follows from the fact that magnetic flux lines are always continuous in the form of closed loops, as was mentioned in Sec.3.4.

In Fig.4.8(a), $\sum\phi_i = \phi$ and $\sum\phi_o = (\phi_1 + \phi_2)$, where $\phi = \phi_1 + \phi_2$. Hence, $\sum\phi_i = \sum\phi_o$. Otherwise, $[(+\phi) + (-\phi_1) + (-\phi_2)] = 0$, or $\sum\phi = 0$. Fig.4.8(b) shows the equivalent electric circuit, wherein $I = I_1 + I_2$ is the analogous current law.

4.7.3 Mmf law of magnetic circuit:

Mmf law in magnetic circuits is analogous to Kirchhoff's voltage law in electric circuits. Tracing through any closed loop in a magnetic circuit along any chosen direction, a rise in magnetic PD is conventionally taken as positive and a fall as negative. Mmf law states that the algebraic sum of the magnetic PDs round any closed loop in a magnetic circuit is zero. Mathematically this law is stated as $\sum V_M = 0$. This law is also stated as that the total rise in magnetic PD along any chosen direction in any closed loop in a magnetic circuit equals the total fall in magnetic PD.

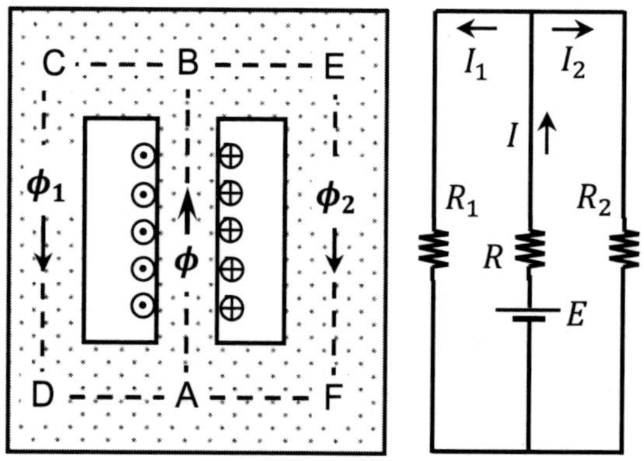

(a) Composite Magnetic Circuit

(b) Equivalent Electric Circuit

Fig. 4.8 Flux law and mmf law explained

Stated mathematically, $\sum V_M(rise) = \sum V_M(fall)$. In the composite magnetic circuit shown at Fig.4.8(a), there are three loops: ABCDA, ABEFA & BCDAFEB. Let the reluctance of the portion AB be S, and those of BCDA and BEFA be S_1 and S_2 respectively. In the loop ABCDA, starting from A, tracing along the direction A>B>C>D and reaching back to A, represented by the notation $\overrightarrow{\text{ABCDA}}$, mmf law applies as $[+NI - \phi.S - \phi_1.S_1 = 0]$. Similarly, in loop $\overrightarrow{\text{ABEFA}}$, we have $[+NI - \phi.S - \phi_2.S_2 = 0]$. On subtracting the latter equation from the former one, we get $[-\phi_1.S_1 + \phi_2.S_2 = 0]$ which corresponds to the loop $\overrightarrow{\text{BCDAFEB}}$. As in Current Electricity, mmf law follows from the fact that the potential at any point in a circuit is unique. For example, if the magnetic potential at the point A is V_M, then starting from A, following any complete loop, say $\overrightarrow{\text{ABCDA}}$, and reaching back to A, the rise and fall in magnetic potential must equal in order to maintain the uniqueness of the potential at the point P at V_M. Hence, we must have $NI = [\phi.S + \phi_1.S_1]$, which was given above by this law.

We now sum up our understanding of electric and magnetic potential, PD, emf and mmf.

1) Electric potential refers to the amount of work done upto a point in an electric field or on the surface of a charged conductor (Electrostatics), or in an electric circuit (Current Electricity). Similarly, magnetic potential refers to the amount of work done upto a point in a magnetic field (Magnetism), or in a magnetic circuit (Electromagnetism). In all cases, however, the potential refers to some zero level.

2) Electric or magnetic PD refers to the amount of work done across a pair of points in electric or magnetic field or circuit.

3) Emf in electric circuit is the source of electric PD and refers to the total amount of work done once round the entire closed circuit, which is the sum of all PDs $(I.R)$ across the various loads in the circuit. Hence, $E = \sum(I.R)$, or $E - \sum(I.R) = 0$.

4) Mmf in magnetic circuit is the source of magnetic PD and refers to the total amount of work done once round the entire closed circuit, which is the sum of all magnetic PDs $(H.l)$ across the various portions of the circuit. Hence,

$F = \sum(H.l)$, or $F - \sum(H.l) = 0$. Since $H.l = \phi.S$, we also have

$F = \sum(\phi.S)$, or $F - \sum(\phi.S) = 0$.

Electrostatics	**Current Electricity**	**Magnetism**	**Electromagnetism**
$V = \int E.dx$	$V = \text{I.R}$ $E = \sum(I.R)$	$V_M = \int H.dl$	$V_M = H.l = \phi.S$ $F = \sum(H.l) = \sum(\phi.S)$
Electric flux exists	Electric current exists	Magnetic flux exists	Magnetic flux exists

4.7.4 Magnetic leakage and fringing: An air gap forms an essential part of many practical magnetic circuits. In Fig.4.9, an instrument coil operates in the air gap of the magnetic circuit of an electromagnet. The air gap flux is thus the *"useful"* flux ϕ_U. The portion of the *"total"* flux ϕ_T produced by the magnetising winding which does not reach the point of use, the air gap in this case, but completes the circuit following shorter air paths, is the *"leakage"* flux ϕ_L. Hence $\phi_T = \phi_U + \phi_L$, and the *leakage factor* is defined as

$$F_L = \frac{\phi_T}{\phi_U} \qquad \text{.......................................} (4.16)$$

which is thus always >1. Any form of leakage is always undesirable. Magnetic leakage is minimised by distributing the winding over the core surface as far as possible, as shown at Fig.4.7. In electrical machines the leakage factor usually lies in the range 1.15 – 1.25. The *"fringing"*, as was discussed in detail in Sec.1.7.2, is nothing but the diverging or bulging out of the useful flux in the air gap. Owing to leakage, flux density along the length of the iron core no longer remains uniform but gradually decreases from its maximum at the winding portion to the minimum at the air gap portion. As it is impossible to allow for such variation, and a higher flux density demands a higher mmf, to estimate a safe value of the mmf requirement it is customary to assume that the maximum flux density continues upto the air gap portion. In other words, it is assumed that the total flux ϕ_T continues through the iron core and passes the air gap, and a portion ϕ_U of the same is useful. Under such assumption, for a value of F_L of 1.2, we get $NI = \phi_T.S = 1.2\,\phi_U.S$, where NI is the mmf provided by the magnetising winding and S is the combined reluctance of the iron core and the air gap in series. This shows that for a given value of $\emptyset_U$, 20% extra mmf is required to compensate for the leakage. Alternatively, a useful mmf equal to $\phi_U.S$ acting on the reluctance S produces the useful flux ϕ_U, plus a 'leakage" mmf equal to $\phi_L.S$ acting on the *same* reluctance S produces the leakage flux ϕ_L. By superimposition, the total flux $\phi_T = \phi_U + \phi_L$, and the total mmf $NI = \phi_U.S + \phi_L.S = \phi_T.S = F_L.\,\phi_U.S$.

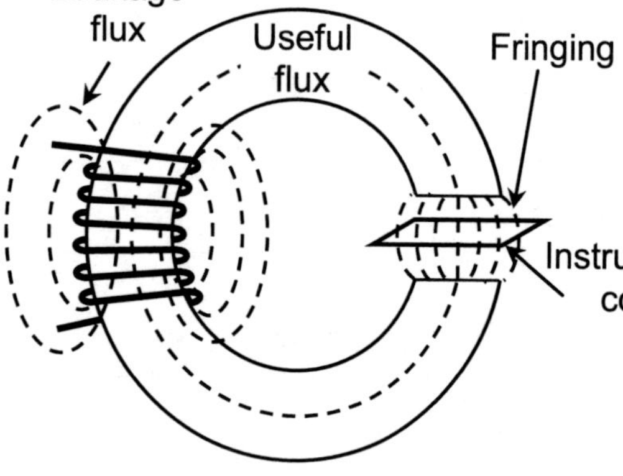

Fig. 4.9 Magnetic leakage and fringing

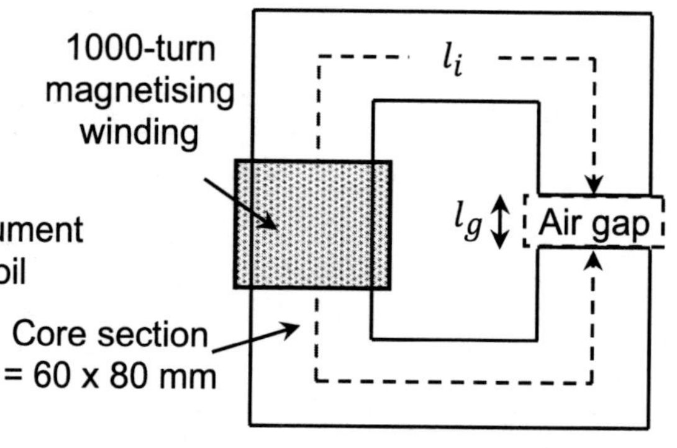

Fig. 4.10 Illustration of magnetic circuit by numerical example

A numerical example will help better understanding of the magnetic circuit.

Example 4.1: *In the magnetic circuit shown at Fig.4.10, all limbs of the iron section measures 60 x 80 mm. Mean iron length is 1.5 m and air gap length is 4 mm. Calculate the current required in the 1000-turn magnetising winding to produce an average flux density of 1.08 T in the air gap. Assume a leakage factor 1.2 and that the average air gap area increases by 2 % due to fringing. The magnetic material is stalloy, having a relative permeability of around 3000 for a flux density of 1.1 T.*

Solution: Useful flux ϕ_U = flux density in iron B_i x iron area A_i

= average flux density in air gap B_g x average air gap area A_g

Therefore, $B_i = \dfrac{B_g.A_g}{A_i}$. Now, $A_g = 1.02\,A_i$.

So, $B_i = \dfrac{B_g.1.02\,A_i}{A_i} = 1.02\,B_g = 1.02\,x\,1.08 = 1.1$ T

Relative permeability of the iron at this flux density of 1.1 T has been given as 3000. Either of the two following methods can now be applied for the solution.

First method: $\boldsymbol{V_M = \phi.S}$

$$\phi_U = B_i\, A_i = 1.1\, x\, 60\, x\, 80\, x\, 10^{-6} = 5.28\, x\, 10^{-3}\ \text{Wb}$$

Reluctance of iron core $S_i = \dfrac{Iron\ length\ l_i}{\mu_0 . \mu_r . A_i} = \dfrac{1.5}{4\pi.10^{-7}\,.\,3000\,.\,60\,x\,80\,x\,10^{-6}}$

$= 8.29\, x\, 10^{4}$ A/Wb

Similarly, reluctance of air gap $S_g = \dfrac{Air\ gap\ length\ l_g}{\mu_0 . A_g} = \dfrac{4.10^{-3}}{4\pi.10^{-7}\,.\,1.02\,.\,60\,x\,80\,x\,10^{-6}}$

$= 65\, x\, 10^{4}$ A/Wb

So the total reluctance of the magnetic circuit $S = S_i + S_g$

$= (8.29 + 65)x\, 10^{4} = 73.29\, x\, 10^{4}$ A/Wb.

Magnetic PD across iron core $V_{M(iron)} = \phi_U \, . \, S_i$

and that across air gap $V_{M(gap)} = \phi_U \, . \, S_g$.

So, mmf required for useful flux

$$V_M = V_{M(iron)} + V_{M(gap)} = \phi_U\left(S_i + S_g\right) = \phi_U . S$$

Mmf required to be provided on the magnetising winding to create the total flux ϕ_T

$$NI = F_L . V_M = 1.2\, \phi_U . S$$
$$= 1.2\, x\, 5.28\, x\, 10^{-3}\, x\, 73.29\, x\, 10^{4} = 4650\ \text{A}.$$

Hence the required current on the 1000-turn winding

$$I = 4650/1000 = 4.65\ \text{A}$$

Second method: $\boldsymbol{V_M = H.l}$

$$V_{M(gap)} = H_g . l_g = \frac{B_g}{\mu_0} l_g = \frac{1.08}{4\pi.10^{-7}} .\, 4.10^{-3} = 3.44\, x\, 10^{3}\ \text{A}$$

$$V_{M(iron)} = H_i . l_i = \frac{B_i}{\mu_0 \mu_r} .\, l_i = \frac{1.1}{4\pi.10^{-7}.3000} .\, 1.5 = 0.44\, x\, 10^{3}\ \text{A}$$

$$V_M = V_{M(iron)} + V_{M(gap)} = 3.88\, x\, 10^{3}\ \text{A}$$

$$NI = F_L . V_M = 1.2\, x\, 3.88\, x\, 10^{3}\, A = 4650\ \text{A}$$

Hence the required current on the 1000-turn winding, $I = 4650/1000 = 4.65$ A

In the second method the core section dimensions are superfluous or redundant data. In this example the second method appears simpler. In fact this method is commonly used in magnetic circuit design problems for mmf calculation. The first method, however, may be simpler in many other problems, and one has to decide which method to adopt when.

Table 4.2: Comparison of Electric and Magnetic Circuits

Electric Circuit		**Magnetic Circuit**	
Analogy			
$I = \frac{E}{R} = \frac{E}{1/G} = EG \quad R = \frac{\rho l}{A}$ $G = \frac{1}{R} = \frac{A}{\rho l} = \frac{\gamma A}{l} \quad \gamma = \frac{1}{\rho}$		$\phi = \frac{F}{S} = \frac{F}{1/P} = FP \quad S = \frac{l}{\mu A}$ $P = \frac{1}{S} = \frac{\mu A}{l}$	
Quantity	Unit	Quantity	Unit
Electromotive Force (E), $E = \sum I.R$	volt (V)	Magnetomotive Force (F), $F = NI = \sum \phi.S$	ampere (A)
Electric Potential or PD (V), $V = IR$	volt (V)	Magnetic Potential or PD (V_M), $V_M = \phi.S$	ampere (A)
Electric Current (I)	ampere (A)	Magnetic Flux (ϕ)	weber (Wb)
Resistance (R), $R = \rho \frac{l}{A}$	ohm (Ω)	Reluctance (S), $S = \frac{l}{\mu A}$	ampere/weber (A/Wb) or per henry (H^{-1})
Resistivity (ρ)	ohm.metre (Ωm)	Reluctivity ($1/\mu$)	metre/henry (m/H)
Conductance(G), G= 1/R	siemens (S)	Permeance (P), P= 1/S	weber/ampere (Wb/A) or henry (H)
Conductivity (γ), γ= 1/ρ	siemens/metre (S/m)	Permeability (μ)	henry/metre (H/m)
Electric circuits obey circuit laws like Kirchhoff's voltage and current laws		Magnetic circuits also obey similar mmf and flux laws.	
Distinction			
Electric circuit formulae are accurate		Magnetic circuit formulae are approximate	
Energy is required to be supplied in maintaining electric current		Energy is not required to be supplied in maintaining magnetic flux	
Electrical Power (P), $P = VI$ Electrical Energy (W), $W = Pt = VIt$		Energy stored in magnetic field (SE_M) $SE_M = \frac{1}{2} v_M \phi = \frac{1}{2} NI\phi$	

4.8 Equivalence between Magnet and Electromagnet: Since an electric current is associated with magnetism, there must exist some equivalence between the two. If a current need be replaced by a magnet of equal magnetism, or the reverse, what should be the relation between the values of the current and the pole strength of the magnet ? This is the question we aim to answer. The equivalence can be established from some alternative considerations. The most simple and convincing equivalence follows directly from our previous discussions in Sec.3.7.2 and Sec.4.5. Thus, for the magnets of Figs.3.6 and 4.6 of identical shape and size to be equivalent, comparison of the corresponding Eqns.3.10 and 4.7 reveal the identity as

$$\frac{J}{\mu} = \frac{NI}{l}, \text{ or } \frac{m}{\mu A} = \frac{NI}{l},$$

where $A = \pi R^2$ is the area of the magnet, or the electromagnet, which are the same. Hence,

$$\mu NI = \frac{ml}{A} = \frac{p_m}{A} = Jl = \textit{Strength of magnetic shell} \quad \text{......... (4.17)}$$

where $p_m = ml$ is the moment or strength of the magnet. This equivalence was first established by Ampere, and is called *Ampere's Theorem*, which states that *the magnetic effect of a closed circuit of NI ampere-turns is the same as that of a magnetic shell occupying the same space as enclosed by the circuit and of strength μNI.*

4.9 Magnetic Potential near Current Loop:

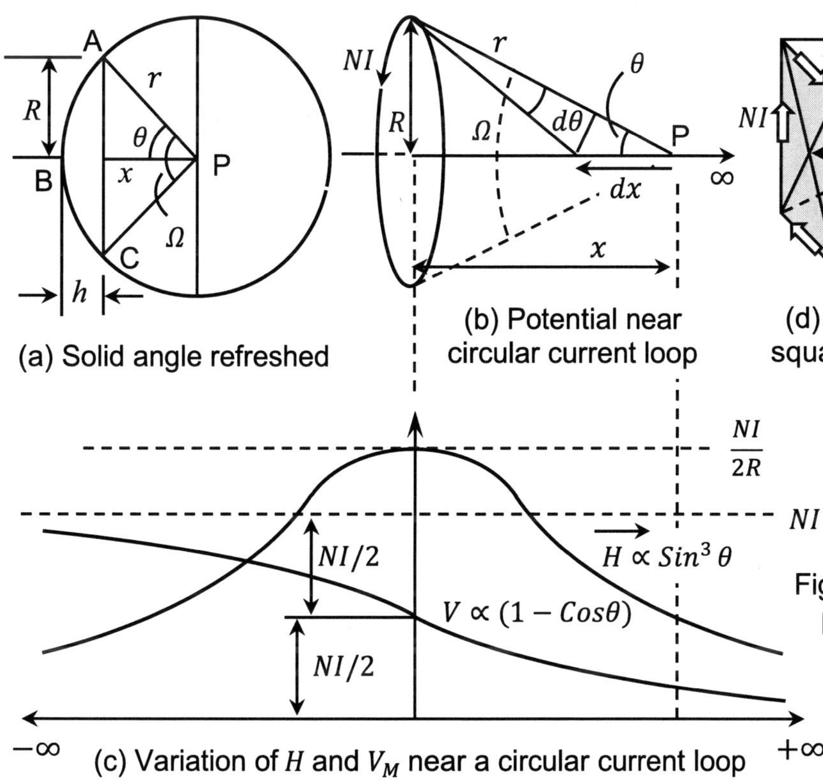

Fig.4.11 Magnetic potential near current loop

Before we discuss this topic, it may be well to refresh our previous knowledge of the solid angle. In Fig.4.11(a), the circle shown is a sphere. Its curved surface ABC is called spherical cap or dome, and its area is given by

$$A = 2\pi rh = 2\pi r(r - x) = 2\pi r(r - rCos\theta) = 2\pi r^2(1 - Cos\theta)$$

The solid angle the cap subtends at the centre P of the sphere

$$\Omega = \frac{A}{r^2} = 2\pi(1 - Cos\theta) \text{ or, } 1 - Cos\theta = \frac{\Omega}{2\pi}$$

This result will be used below.

In Fig.4.11(b), the field strength at point P on the axis of and distant x from a circular current loop of radius R and carrying NI ampere-turns, by Eqn.4.5, $H = \left[\frac{NI}{2R}\right] Sin^3\,\theta$. Hence, magnetic potential at point P, the work done by some external agency in moving a test pole from infinite distance to the point,

$$V_M = -\int_{\infty}^{x} H.dx = -\frac{NI}{2R}\int_{\infty}^{x} Sin^3\,\theta\,dx$$

In Fig.4.11(b), $dx\,Sin\,\theta = r.d\theta$ and $Sin\,\theta = R/r$. Multiplying these identities,

$$Sin^2\,\theta\,dx = R.d\theta$$

As θ increases with decrease in x, this becomes $-Sin^2\,\theta\,dx = R.d\theta$. Hence,

$$V_M = \frac{NI}{2}\int_{0}^{\theta} Sin\,\theta\,d\theta$$

$$= \frac{NI}{2}[-Cos\,\theta]_0^{\theta}$$

$$= \frac{NI}{2}(1 - Cos\,\theta) = \frac{NI}{2}\frac{\Omega}{2\pi} = \frac{NI}{4\pi}\Omega \qquad \text{......................... (4.18)}$$

where Ω is the solid angle subtended by the current loop at point P.

At the centre of the current loop, $\theta = \pi/2$, $Cos\theta = 0$, $\Omega = 2\pi$, so $V_M = NI/2$. At infinite distance, $\theta = 0$, $Cos\theta = 1$, $\Omega = 0$, so $V_M = 0$. The variations of H and V_M with the distance x have been plotted in Fig.4.11(c). In these curves, $\theta = 0$ at $x = +\infty$ and $\theta = \pi$ at $x = -\infty$. Thus, the work done against the field in moving a test pole from $+\infty$ to $-\infty$ is NI, which is the mmf of the current loop. It may now be appreciated that the solenoidal H-curve of Fig.4.6(b) is development of the present circular-loop H-curve with central portion stretched along X- axis.

Another point follows from Eqn.4.18. Appearance of the solid angle in this equation implies its general validity. Thus, the equation remains valid (a) for all points in space, even if outside the axis of the current loop, and (b) for any geometry of the current loop, even if the loop is not a planar. For example, in Fig.4.11(d), the potential at the centre of the cube (with six equal surfaces symmetrically surrounding the centre point),

$$V_M = \frac{NI}{4\pi}\frac{4\pi}{6} = NI/6$$

One more point needs mention here. In Fig.4.11(b), as the magnetic pole approaches the current loop, the current I will get modified due to electromagnetic induction which we ignored in this chapter. The complete subject will be understood in next chapter (Sec.5.7).

4.10 Magnetic Potential on the Axis of Solenoid: Comparing Figs.3.6 and 4.6, magnetic potential on the axis of a solenoid can simply be derived by equivalence, Sec.4.8. Thus, substituting NI/l for J/μ from Eqn.4.17 in Eqn.3.18, the potential at point P in Fig.4.6 becomes

$$V_M = \frac{NI}{2l}\left[\sqrt{[x-(l/2)]^2+R^2} - \sqrt{[x+(l/2)]^2+R^2} + l)\right] \quad \text{.......... (4.19)}$$

This can also be derived otherwise. In Fig.4.6, potential at P due to the elemental length dy of the solenoid, by Eqn.4.18, $dV_M = \frac{nI.dy}{2}(1 - Cos\,\theta)$. As the elemental circular loop moves along right or comes closer to point P, y decreases, θ increases, $Cos\,\theta$ decreases and so dV_M increases. Hence, the potential at P due to the entire solenoid is given by

$$V_M = -\int_{y=y_1}^{y=y_2} dV_M = -\frac{NI}{2l}\int_{y_1}^{y_2}(1 - Cos\,\theta)dy = \frac{NI}{2l}\int_{y_1}^{y_2}(Cos\,\theta - 1)dy$$

$$= \frac{NI}{2l}\int_{y_1}^{y_2}\left(\frac{y}{\sqrt{y^2+R^2}} - 1\right)dy$$

$$= \frac{NI}{2l}\left[\sqrt{y^2+R^2} - y\right]_{y_1}^{y_2}$$

$$= \frac{NI}{2l}\left[\sqrt{\{x-(l/2)\}^2+R^2} - \{x-(l/2)\}\right] - \frac{NI}{2l}\left[\sqrt{\{x+(l/2)\}^2+R^2} - \{x+(l/2)\}\right]$$

$$= \frac{NI}{2l}\left[\sqrt{[x-(l/2)]^2+R^2} - \sqrt{[x+(l/2)]^2+R^2} + l)\right],$$

which is Eqn.4.19.

4.11 Potential Energy of Magnetic Pole near Current Loop: Potential energy of a magnetic pole m at a point of magnetic potential V_M, Eqn.3.14, $PE_M = m.V_M$. When the magnetic field is due to a current loop of NI ampere-turns, it follows from Eqn.4.18

$$PE_M = m.\frac{NI}{4\pi}\Omega = NI\left[\frac{m\Omega}{4\pi}\right]$$

Referring to Fig.4.11(b), the quantity $m\Omega/(4\pi)$ is the portion ϕ of the total flux originating from the pole m lying at the point P, that threads through the current loop, or $\phi = m\Omega/(4\pi)$. Hence,

$$PE_M = NI\phi \quad \text{............................ (4.20)}$$

The ϕ's are different in Eqns.4.15 and 4.20. It should be noticed that the PE_M possessed by the pole is with respect to zero reference at infinite distance, and also that ϕ is *opposing flux* (magnetically opposes the current loop). As reaction, this is also the PE_M of the current loop in the field of the magnetic pole. These observations will be found useful in Sec.4.17.

Part B: Reaction of Magnet on Electric Current

In Part A we studied topics related to the fact that "electric current exerts force on magnet". It follows from Newton's Third Law of Motion that the magnet also exerts an equal and opposite force on the current. This we call the reaction of magnet on electric current.

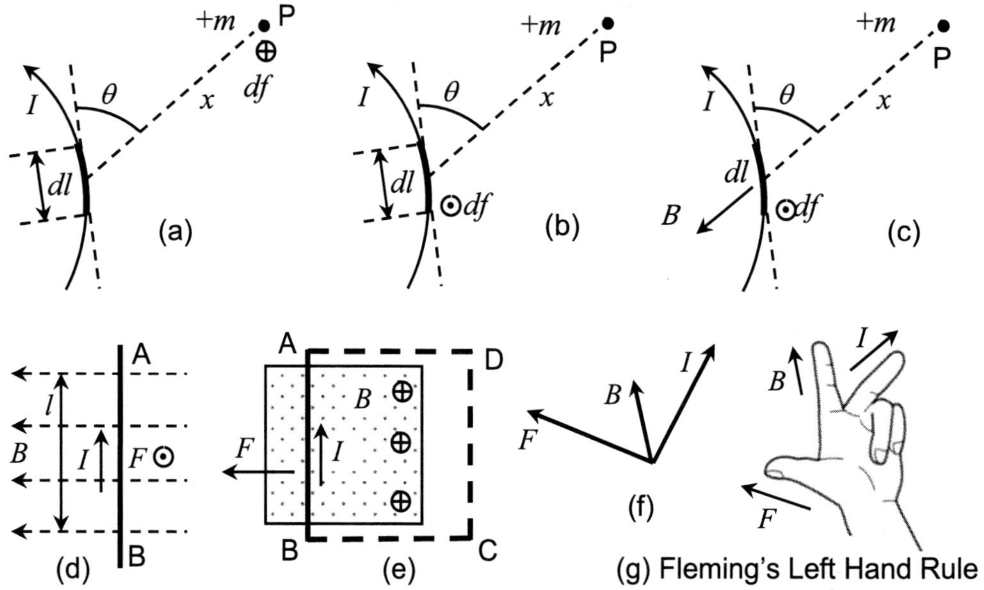

Fig.4.12 Force on current carrying conductor and coil in magnetic field

4.12 Force on Current Carrying Conductor or Coil in Magnetic Field: Reproducing Fig.4.1 here as Fig.4.12(a), the force exerted by the elemental length dl of the current I on the magnetic pole m is given by Biot-Savart-Laplace Law,

$$df = \frac{m.I.dl.Sin\,\theta}{4\pi x^2} \text{ (Eqn.4.1)},$$

and the direction of this force is perpendicular to and into the plane of the paper, as shown by the encircled cross. An equal and opposite force df exerted by the magnetic pole m on the elemental length dl of the current I is therefore perpendicular to and away from the plane of the paper, as shown by the encircled dot in Fig.4.12(b). Now, the flux density at dl due to the pole m being

$$B = \frac{m}{4\pi x^2}, \text{ Fig.(c)},$$

we have $df = B.I.dl.Sin\,\theta$. If the conductor AB is straight and its *active* length l lies in a uniform flux density B (as in a narrow air gap of a magnet), and they are in perpendicular orientation as shown at Fig.(d), so that $Sin\,\theta = Sin\,90 = 1$, the force exerted by the magnetic field B on the length l of the current I becomes

$$F = BIl \qquad \text{.. (4.21)}$$

The directions of B, I and F in Fig.(d) are oriented along the three mutually perpendicular three-dimensional X-Y-Z axes in space, and can, for convenience, be reoriented as in Fig.(e), wherein the flux density B directs perpendicular to and into the plane of the paper over the dotted rectangular field area. These directions can be represented by three vectors shown at Fig.(f), and best remembered by *Fleming's Left Hand Rule* depicted by Fig.(g).

Whereas the force exerted by a magnet on a current (carrying conductor) is due to the interaction between the lines of force of the two elements, it may be conceived that the current does not possess any lines of force to modify the field of the magnet but experiences the force by virtue of its presence in the field of the magnet. As such, the current may be treated as a *test current* of strength I, as was indicated in Sec.1.3.

The conductor AB in Fig.4.12(d) must be a portion of some circuit or coil like ABCDA, Fig.(e). Thus, if this circuit is of N turns, the force on the current carrying coil

$$F = BNIl \quad \text{...} \quad (4.22)$$

As electric motors operate on this principle '*a current-carrying conductor lying in a magnetic field experiences a force*', we usually refer it as '*motor principle*'. Fleming's left hand rule that gives the direction of such force, with respect to the directions of the magnetic field and the electric current, is called the '*motor rule*'. Sir John Ambrose Fleming (1849-1945), FRS, an English electrical engineer and physicist, formulated the left hand (motor) rule, and also the right hand (generator) rule (Sec.5.7.2), in the late 19th.Century.

4.13 Magnetic Moment of Electromagnet: Magnetic moment of an electromagnet of NI ampere turns, may it be a current loop or a solenoid, can be derived simply from Sec.3.9 and the magnet to electromagnet equivalence, Eqn.4.17, as $p_m = \mu NIA$, A being the area of the electromagnet. This definition was used in pre-1930's text books. In later textbooks, it is defined as another vector $\boldsymbol{\mu}$. As μ is popular as permeability, we prefer to use the symbol p_c for the magnetic moment of an electromagnet, or the *electromagnetic moment of a coil*, the suffix c representing *current*. It is defined as

$$p_c = NIA \quad \text{..................} \quad (4.23)$$

which thus differs from p_m by μ times. Fig.4.13 distinguishes these definitions pictorially.

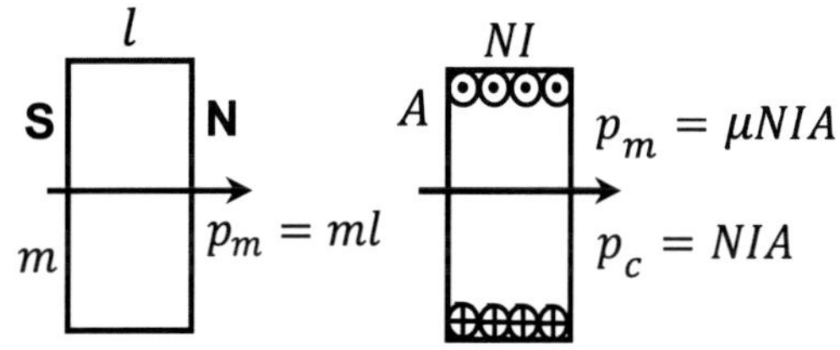

Fig.4.13 Magnetic moment of magnet and electromagnet

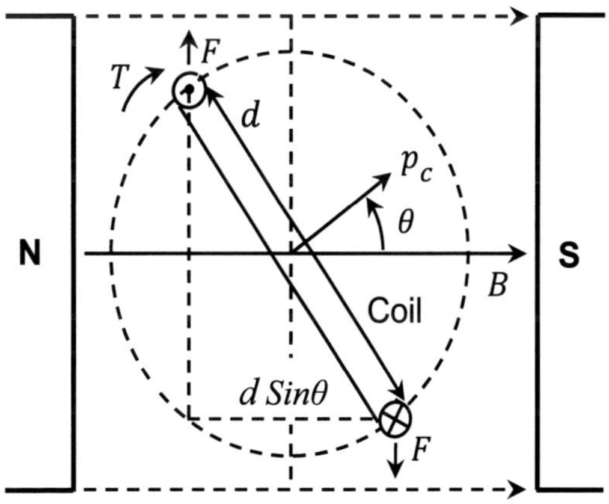

Fig.4.14 Torque on current carrying coil in magnetic field

4.14 Torque on Current Carrying Coil in Magnetic Field: In Fig.4.14, the rectangular coil of width d and carrying NI ampere-turns lies in a uniform magnetic field of flux density B directing from left to right. The coil is capable of rotation about an axis passing through its centre and normal to the plane of the paper, as shown. The active length l of the coil within the magnetic field lies normal to the plane of the paper.

When the magnetic axis of the coil is displaced from the magnetic field by an angle θ, as shown, the two equal and opposite forces $F = BNIl$ on the two active coil sides are $dSin\theta$ apart.

This constitutes a couple, and the resulting clockwise torque on the coil,

$$T = F.dSin\theta = BNIld.Sin\theta$$

$$= BNIA.Sin\theta \qquad \text{(Coil area } A = l.d) \qquad \text{.................... (4.24)}$$

The quantity NIA is its electromagnetic moment p_c. In the general case of any geometry of the coil section, therefore, the torque is given by

$$T = p_c.B.Sin\theta \qquad \text{.................................. (4.25)}$$

This torque tends to align the coil axis along the magnetic field to which it is due.

4.15 Practical Considerations of Force and Torque in Electrical Equipment: Practical situations, although operate on the principles discussed above, differ in physical arrangement. We illustrate below two such situations.

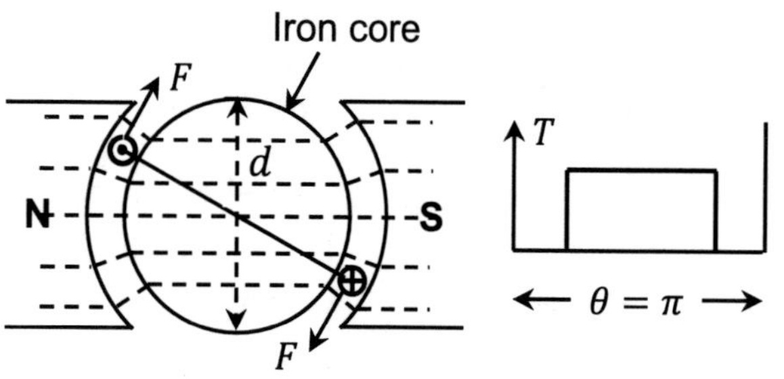

Fig.4.15 Arrangement in Permanent Magnet Moving Coil instrument

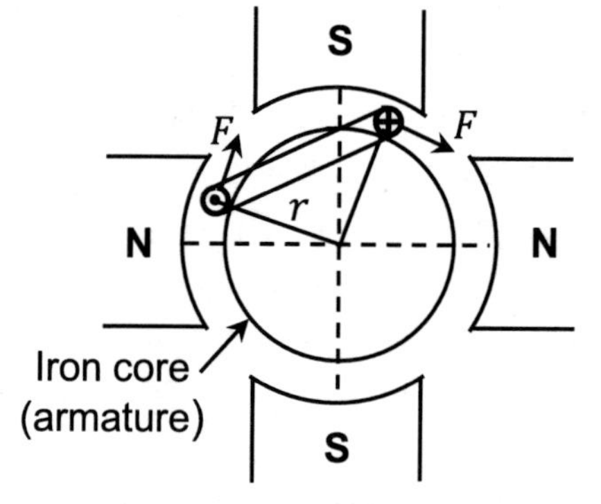

Fig.4.16 Arrangement in 4-pole DC Motor

1) If the pole faces of the magnet shown in Fig.4.14 are of circular profile and a cylindrical iron core be inserted in the air gap, the modified arrangement, Fig.4.15, prevails in the Permanent Magnet Moving Coil (PMMC) instrument. The iron core is stationary and the coil is spring-controlled to rotate in the air gap within the magnetic area of the magnet. The iron core makes the air gap flux radial, and so the forces on the coil sides tangential. As such, the term $Sin\theta$ in Eqn.4.24 does not appear, and we *blindly* write the torque as $BlINd$. The force or torque is thus constant (independent of the position of the coil) within the operating angle of the instrument.

2) Fig.4.16 shows the arrangement in a 4-pole DC machine. In this case, the iron core (armature) houses the coils in its surface grooves (slots), and the assembly rotates. The moment of the force F on one coil side is $F.r$, and the torque on the coil is $2F.r = F.d$. A continuous belt of such multi-turn coils (not shown in the figure) cover the entire surface of the armature. The torque is given by the summation of all such torques. $T = \sum(F.d) = \sum(BlINd)$. As any coil side passes from one pole to next opposite pole, the current through the coil is reversed by the "commutator", so that the direction of the force or torque remains unaltered. Hence the force or torque is again constant.

Part C: Interaction Between Electric Currents

As an electric current is associated with magnetic field, a force exists between current carrying conductors or circuits. The magnitude and direction of such force depend on factors like the strength of the currents, size geometry spacing and relative position of the conductors or circuits. Let us consider first the simplest case of the force between two straight and parallel current carrying conductors.

4.16 Force Between Parallel Currents: Fig.4.17 shows two parallel current carrying conductors. If the length of the conductors is very large compared to their distant apart d, the magnetic field at point B due to the current I_1, as was derived in Sec.4.3 (Eqn.4.3), $H = I_1/(2\pi d)$, directed into the plain of the paper. The flux density at B, also directed into the plain of the paper, $B = \mu H = \mu I_1/(2\pi d)$. If we consider a length l of the two conductors, length l of the bottom conductor lies in the magnetic field B. Hence the force on length l of the bottom conductor due to the top conductor,

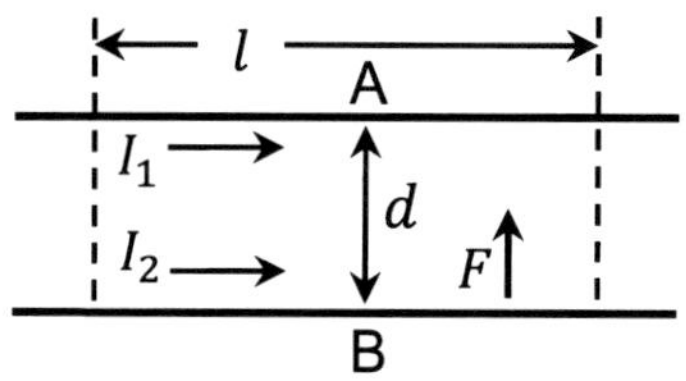

Fig.4.17 Force between parallel currents

$$F = B.I_2.l = \frac{\mu I_1 I_2 l}{2\pi d} \quad \text{.................................... (4.26)}$$

Flemings Left Hand Rule shows that this force is upwards. Hence the force between two conductors carrying currents in the same direction is attractive. If the two currents are in opposite directions, this force will be repulsive one. These directions of the force can be explained by Fig.4.2 in terms of the properties of lines of force. The size or definition of the unit 'ampere' is derived from this equation. If I_1, I_2, l and d are all unity and $\mu = \mu_0$, we get a force $F =$ 2 x 10^{-7}. *Hence the ampere is defined as the current which, if maintained in two thin parallel conductors of infinite length and spaced one metre apart in vacuum, produces a force of 2 x10^{-7} newton between the conductors per metre length of the conductors.*

4.17 Force and Potential Energy of Current Carrying Coils: The force F between, and hence the potential energy PE_M of, neighbouring current carrying coils depend on their relative positions. Following situations may arise. The relative displacement between the coils may be (a) axial, (b) radial (traverse), (c) angular, and of course, any combination of these. Axial and radial displacements involve linear motion, and angular displacement involves circular motion. We consider these situations separately.

4.17.1 Force and PE of coils in axial displacement: Before we discuss this topic, it is necessary to develope a preparatory concept on the flux distribution of two neighbouring current carrying (magnetically or mutually "coupled') coils. When two coils lie in proximity, and any of them is energised, out of the total flux produced, a portion passes through the same coil only and the remaining portion through both the coils. The former portion is *self* or *leakage flux* and the latter *mutual flux*. Thus, in Fig.4.18(a), when coil 1 is energised *alone*,

Total flux ϕ_1 = Leakage flux ϕ_{1L} + Mutual flux ϕ_{12}

$$\text{or}\quad \frac{N_1I_1}{S_1} = \frac{N_1I_1}{S_{1L}} + \frac{N_1I_1}{S}$$

Where, S_1 = Reluctance of the magnetic circuit of coil 1 (for total flux ϕ_1)

S_{1L} = Reluctance of the leakage magnetic circuit of coil 1 (for flux ϕ_{1L}), and

S = Reluctance of the common magnetic circuit through both the coils (for flux ϕ_{12})

And, $$\frac{1}{S_1} = \frac{1}{S} + \frac{1}{S_{1L}} \qquad \text{.. (4.27)}$$

as S_1 comprises of parallel combination of S and S_{1L}.

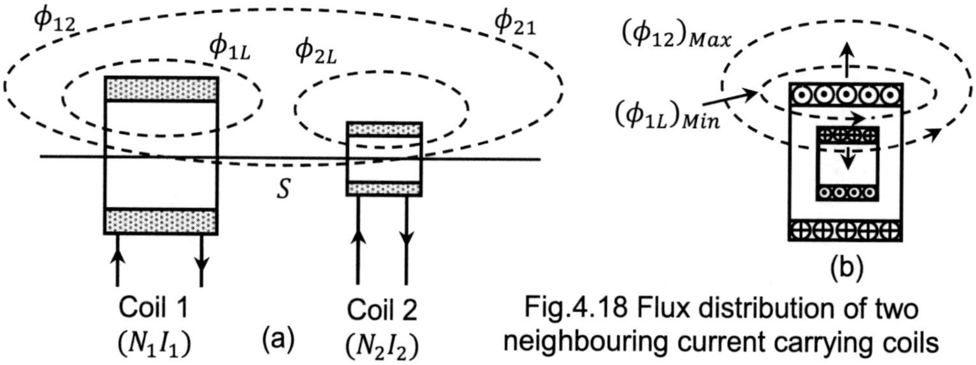

Fig.4.18 Flux distribution of two neighbouring current carrying coils

Similarly, when coil 2 is energised *alone*, by similar reasoning and using similar nomenclature,

$$\phi_2 = \phi_{2L} + \phi_{21},\quad \frac{N_2I_2}{S_2} = \frac{N_2I_2}{S_{2L}} + \frac{N_2I_2}{S},$$

And, $$\frac{1}{S_2} = \frac{1}{S} + \frac{1}{S_{2L}} \qquad \text{.. (4.28)}$$

Further, the relations $\phi_{12} = \frac{N_1I_1}{S}$ and $\phi_{21} = \frac{N_2I_2}{S}$ provide $S = \frac{N_1I_1}{\phi_{12}} = \frac{N_2I_2}{\phi_{21}}$,

$$\text{or, } N_1I_1\phi_{21} = N_2I_2\phi_{12} \qquad \text{.. (4.29)}$$

When the coils change their relative positions, S_1 and S_2 do not change. If the coils come closer, for example, S decreases, and both S_{1L} and S_{2L} increase, but their values are so adjusted that S_1 and S_2, as given by Eqns.4.27 and 4.28 respectively, remain unaltered. When both the coils are energised *together*, the total or net mutual flux that links both the coils, $\phi = \phi_{12} \pm \phi_{21}$, for the magnetic fields of the two coils acting in conjunction or in opposition respectively. This preparatory concept will be utilised in the present topic, as well as in next chapter on Electromagnetic Induction.

In line with our previous concept (e.g., in Sec.4.12), let us suppose next, coil 2 does not have any magnetic field but experiences the force by virtue of its presence in the field of coil 1, so that $\phi_1 = \phi_{1L} + \phi_{12}$ are the only fluxes in the system. Whereas ϕ_1 is constant, its division varies with axial distance between the coils. In Fig.4.19(d), curve 5-12-1 shows how ϕ_{12} varies with axial distance.

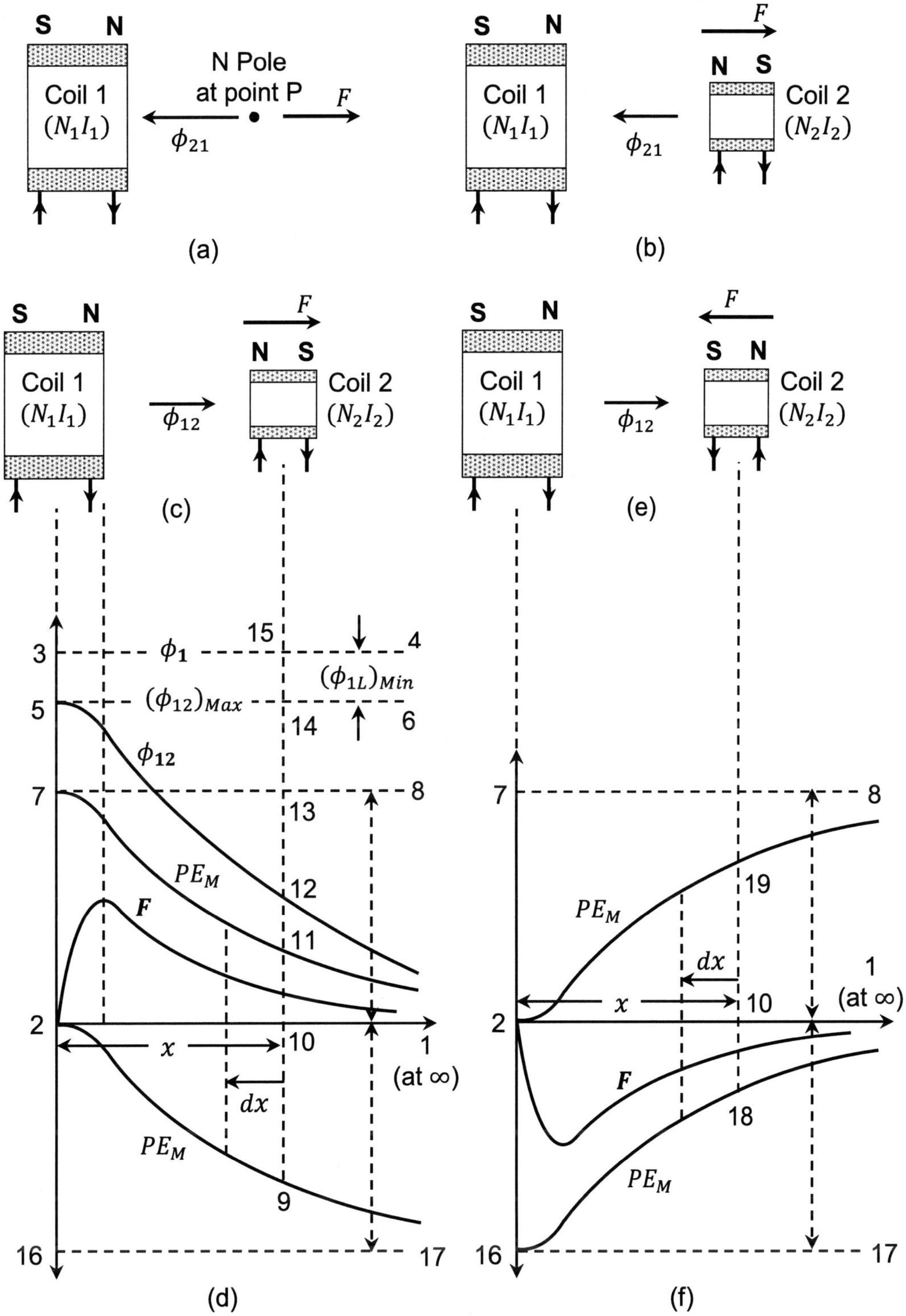

Fig.4.19 Force and PE of coils in axial displacement

If H be the field strength at point P due to coil 1, the flux density at P is $B = \mu H$, and the flux through coil 2, of some small area A, lying at point P, $\phi_{12} = BA = \mu HA$. Thus, $\phi_{12} \propto H$. Hence, the ϕ_{12} curve 5-12-1 is essentially the H- curve shown in Fig.4.6(b). At distance x from the centre of coil 1, ϕ_{12} is height 10-12, and ϕ_{1L} is height 12-15. As the coils come closer, ϕ_{12} increases and ϕ_{1L} decreases by the same amount, so that their total ϕ_1 remains unaltered. This means,

$$\Delta\phi_{12} = -\Delta\phi_{1L} \qquad \text{..............................} \ (4.30)$$

When the coils are centered, Fig.4.18(b), ϕ_{12} attains its maximum value $(\phi_{12})_{Max}$ and ϕ_{1L} attains its minimum value $(\phi_{1L})_{Mim}$. Hence,

$$\phi_1 = \phi_{12} + \phi_{1L} = (\phi_{12})_{Max} + (\phi_{1L})_{Mim}$$

If coil 2 happens to be of larger size than coil 1 (as will be considered in Fig.4.21), $(\phi_{1L})_{Mim}$ will be zero, and as such,

$$\phi_1[= (\phi_{12})_{Max}] = \phi_{12} + \phi_{1L}.$$

As was derived in Sec.4.11, the PE_M of the north pole in Fig.4.19(a), $PE_M = N_1 I_1 \phi_{21}$, where ϕ_{21} is opposing flux. Instead of the north pole, if the flux ϕ_{21} is produced by coil 2, Fig.4.19(b), using Eqn.4.29 we may write that the PE_M of coil 2 in the opposing or repulsive field of coil 1, with respect to zero reference at infinite distance,

$$PE_M = N_2 I_2 \phi_{12} \qquad \text{..............................} \ (4.31(a))$$

corresponding to Fig.4.19(c). This equation is thus an alternative form of Eqn.4.20. As the axial displacement between the coils changes, say, coil 1 fixed and coil 2 moves, the variation of PE_M with the centre distance x between the coils is shown by the curve 7-11-1 in Fig.4.19(d). It is obvious that this PE_M curve is essentially the same as ϕ_{12} curve, differing only in scale. One important point must be noted in this context. As the magnetic force on a point pole m is $F = mH \propto H$, the pattern of the F- curve on a *point pole* near a solenoid, Fig.4.6, is identical to that of the PE_M- curve of a *current loop*, Fig.4.19(d). This apparent ambiguity owes to the distinction between the *isolated pole theory* and *current-ring theory of magnetism*, and is simply resolved by the fact that a current loop is associated with two equal and opposite poles.

The PE_M $(= N_2 I_2 \phi_{12})$ of coil 2 at distance x is given by the height 10-11 in Fig.4.19(d).

The PE_M of coil 2 at distance x with respect to zero reference at zero distance between the coils ($x = 0$), is given by the depth 10-9 on curve 2-9-17, as

$$PE_M = -N_2 I_2[(\phi_{12})_{Max} - \phi_{12}] \qquad \text{.....................} \ (4.31(b))$$

This is negative. If coil 2 comes closer to coil 1 by distance dx, that is, if x decreases by dx, ϕ_{12} increases by $d\varphi_{12}$, so $d\varphi_{12}/dx$ is negative, and whatever zero reference we consider, the PE_M of coil 2 increases by dPE_M, so dPE_M/dx is also negative. In mathematical terms,

$$\frac{dPE_M}{-dx} = N_2 I_2 \frac{d\varphi_{12}}{-dx}$$

and the axial force between the coils, as plotted in Fig.4.19(d),

$$F = -\frac{dPE_M}{dx} = -N_2 I_2 \frac{d\varphi_{12}}{dx} \qquad \text{...........................} \ (4.32)$$

is positive (F acts along positive direction of x), as $d\varphi_{12}/dx$ is negative.

Interestingly, $F = 0$ at $x = 0$. The pattern of the F- curve can also be explained logically. The force between the coils is the resultant of four forces between the two north and two south poles created on the two coils. The situation is analogous to the force between two polarized electrified conductors, Sec.1.13. As the coils come closer, F first increases to some peak, and then falls off sharply as the smaller coil starts entering the larger coil. When the coils are centered ($x = 0$), a balance is established between the four forces, leading to $F = 0$. Thus, if a suspended (friction-less) soft iron rod is inserted into a current-carrying solenoid, the rod will be centered into the solenoid.

If the direction of the current through any one coil, say coil 2, be reversed, as in Fig.4.19(e), the coils will magnetically act in conjunction and ϕ_{12} will be *helping* flux. The PE_M of coil 2, under this condition, can be obtained by replacing I_2 in Eqns.4.31 by $-I_2$. Thus, the PE_M at distance x with respect to zero reference at infinite distance,

$$PE_M = -N_2 I_2 \phi_{12} \quad \text{.............................. (4.33(a))}$$

is negative, depth 10-18 on curve 1-18-16 in Fig.4.19(f). This PE_M curve is simply the mirror image of the curve 1-11-7 in Fig.4.19(d) about the X-axis. The PE_M of coil 2 at distance x with respect to zero reference at zero distance between the coils ($x = 0$),

$$PE_M = N_2 I_2 [(\phi_{12})_{Max} - \phi_{12}] \quad \text{.................... (4.33(b))}$$

is, however, positive, height 10-19 on curve 2-19-8. In this case, as x decreases, PE_M also decreases, so dPE_M/dx is positive. The axial force F between the coils

$$F = -\frac{dPE_M}{dx} = N_2 I_2 \frac{d\varphi_{12}}{dx} \quad \text{.......................... (4.34)}$$

is negative (acts along negative direction of x), as $d\varphi_{12}/dx$ is negative. This latter situation is similar to the attractive force between the plates of a charged capacitor, Sec.1.8.5.

Leakage fluxes exert destructive forces on neighbouring current carrying coils, and this can be understood from the following consideration. Let us first consider the magnetic field of a single solenoid, Fig.4.20. The self flux of the solenoid exerts forces on its own current carrying turns. The directions of such forces, as viewed by Fleming's left hand rule, are shown in the figure. It can be seen that these forces tend to expand the middle turns of the solenoid in radial direction, and squeeze the end turns in axial direction. Self flux thus creates tensile stresses in the winding wire as well as on the winding insulation. To protect against winding deformation and insulation failure, adequate bracing and clamping of windings become necessary in large machines and strong electromagnets.

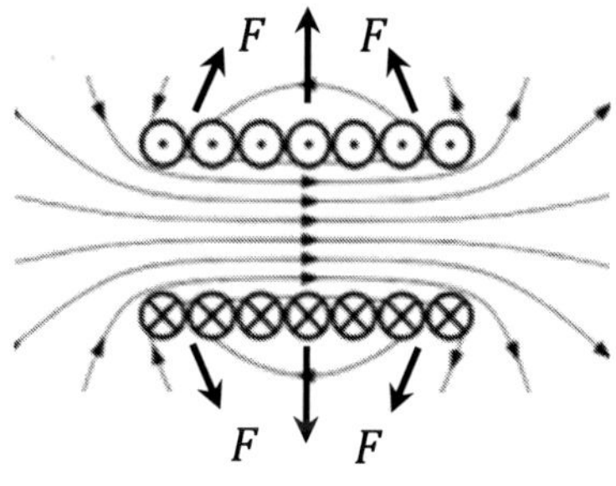

Fig.4.20 Self forces on solenoid turns

Let us now consider the effect of the leakage fluxes in Fig.4.18(b). As discussed above, in addition to their axial squeezing effect, both the mutual and self fluxes tend to radially expand the outer coil and compress the inner coil. If the coils are wound on an iron core, the mutual flux remains confined within it, so will not be able to exert these forces. Only the leakage flux, which is also the minimum when the coils are centered, or when one coil is wound on another, is responsible for these destructive forces. In practice, this situation prevails in transformers. Under normal operation of transformers, these forces are not enough strong to cause any damage to the windings, but during supply of fault current they may do so. One more tendency of the radial forces is to deform rectangular coils into circular shape. Hence rectangular transformer windings are more vulnerable to the radial forces than the circular windings. Some transformer designers strongly recommend circular windings above 5 MVA machines.

If the two coils in Fig.4.18(b) happen to act in conjunction, leakage flux would tend to expand both the coils in radial direction, but such a situation does not prevail in transformers.

Another point follows. If the two axially displaced coils in Fig.4.18(a) are in air, the leakage fluxes, apart from their axial and radial forces on the coils, exert axial attraction when the coils act in conjunction, or repulsion when the they act in opposition. But if the coils are on a common iron core, leakage flux will be almost zero, and no force will be there between the coils. Zero leakage flux means $\phi_{1L} = 0$, or $\phi_{12} = \phi_1 =$ constant, so $d\varphi_{12}/dx = 0$, and hence $F = 0$. In terms of mutual inductance, $dM/dx = 0$ in Eqn.5.23 and hence $F = 0$.

4.17.2 Force and PE of coils in radial (traverse) displacement: In Fig.4.21(a), the force on the coil, $F = BN_2I_2l$, is constant, and positive, being along positive direction of x. The field pushes the coil to the right, and they act in (magnetic) opposition. As the field does not exist beyond position AA' of the coil (at right end of the field, where the force is zero), this position of the coil is its *equivalent zero reference* of PE_M. Against this zero reference, the PE_M of the coil at its BB' position, $PE_M = (-F)(-x) = F.x = BN_2I_2lx$ is positive, height 1-3. As the force is constant, the PE_M line 0-3-4 is a straight line. The term Blx being the opposing flux ϕ_{12} of the field (1) threading through the coil (2), we get $\quad PE_M = N_2I_2\phi_{12}$, which is Eqn.4.31(a).

If the direction of I_2 is reversed, Fig.4.21(b), F acts along negative direction of x, so is negative, $F = -BN_2I_2l$. Here, $-dPE_M = F(-dx)$, or $dPE_M = (-F)(-dx)$. As such, the PE_M of the coil at its BB' position,

$$PE_M = (-F)(-x) = BN_2I_2l(-x) = -BN_2I_2lx$$

$$= -N_2I_2\phi_{12}, \text{ which is Eqn.4.33(a).}$$

This is negative, depth 1-3 on the PE_M line 0-3-4. Here, ϕ_{12} is helping flux.

Situations in Figs.4.21(a) and (b) correspond to zero reference at infinite distance. In practice, as was shown in Figs.4.15 and 4.16, the coil width d is the *pole pitch* (centre distance between two neighbouring opposite pole faces), and as such, Fig.4.21(a) transforms to Fig.(c) wherein the zero reference can be seen to occur when the coil is in

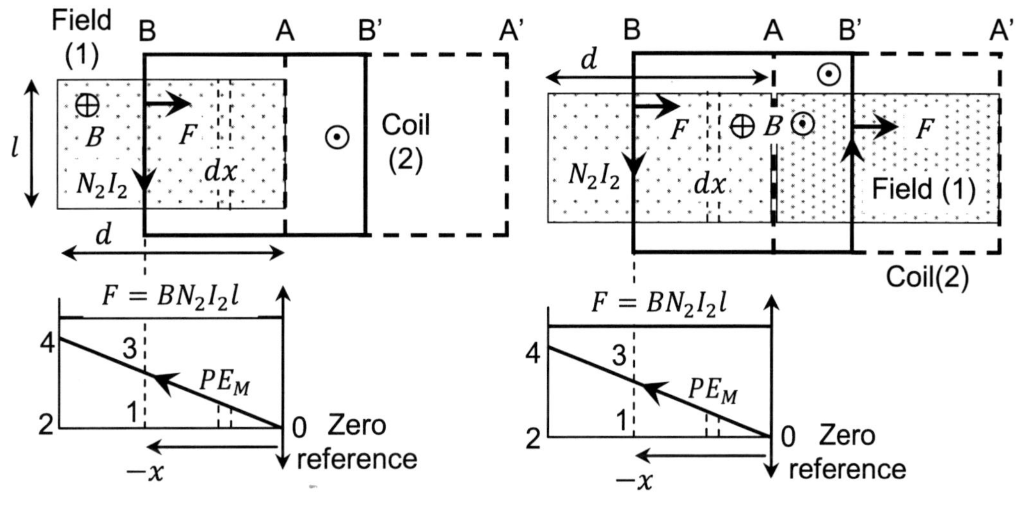

(a) $PE_M = (-F)(-x) = F.x$
$= BN_2I_2lx = N_2I_2\phi_{12}$
($Blx = \phi_{12}$ is opposing flux)

(c) $PE_M = 2(-F)(-x) = 2F.x$
$= 2BN_2I_2lx = N_2I_2.\Delta\phi_{12}$
($2Blx = \Delta\phi_{12}$ is increase in opposing flux)

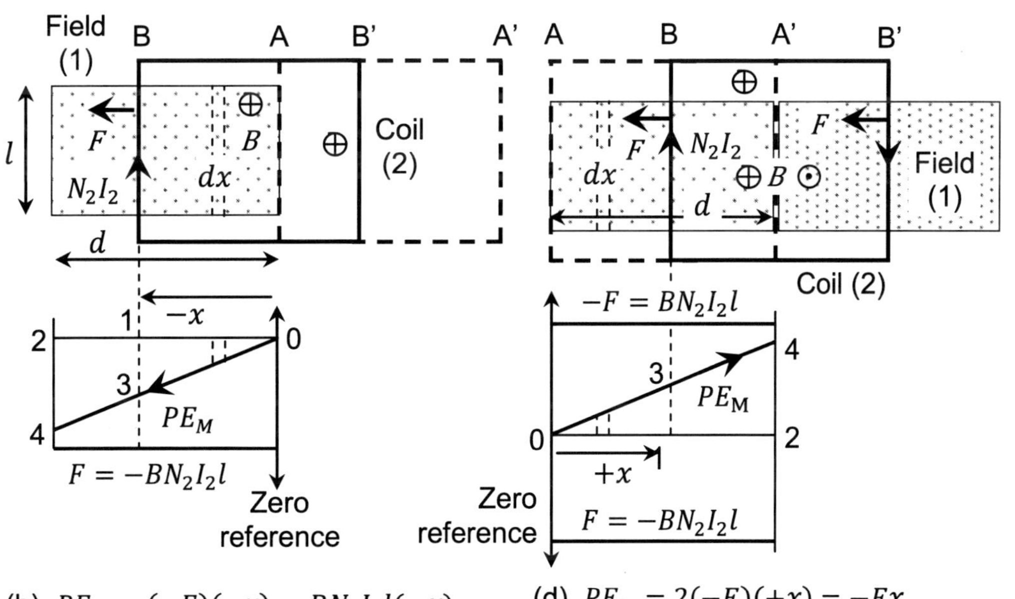

(b) $PE_M = (-F)(-x) = BN_2I_2l(-x)$
$= -BN_2I_2lx = -N_2I_2\phi_{12}$
($Blx = \phi_{12}$ is helping flux)

(d) $PE_M = 2(-F)(+x) = -Fx$
$= 2BN_2I_2lx = N_2I_2.\Delta\phi_{12}$
($2Blx = \Delta\phi_{12}$ is increase in opposing flux)

Fig.4.21 Force and PE of coils in radial (traverse) displacement cumulative alignment with the field. In fact, Fig.(c) is the linear or developed model of the practical situations of Figs.4.15 and 4.16. The PE_M of the coil at its BB' position, $PE_M = 2(-F)(-x) = 2F.x = 2BN_2I_2lx$. As regards the opposing flux, it, at zero reference position AA' of the coil, is not zero now, but $(-Bld)$, and the same at BB' position is $[+Blx - Bl(d-x)] = (2Blx - Bld)$. Hence, the increase in opposition flux as the coil moves from AA' position to BB' position,

$$\Delta\phi_{12} = (2Blx - Bld) - (-Bld) = 2Blx, \text{ and}$$

$$PE_M = N_2I_2.\Delta\phi_{12} \qquad \text{..............................} (4.35)$$

Eqn.4.35 is thus generalised form of Eqn.4.31(a).

Fig.4.21(c) may also be drawn as in Fig.4.21(d), with zero reference at the left. Here the force is negative, $F = -BN_2I_2l$, same as in Fig.(b). Against this zero reference, $dPE_M = -2F.dx$, and the PE_M of the coil at its BB' position, $PE_M = -2F.x = 2BN_2I_2lx$ is positive. The increase in opposition flux $\Delta\phi_{12} = 2Blx$ remains the same as in Fig.4.21(c), so $PE_M = N_2I_2.\Delta\phi_{12}$ also remains the same.

4.17.3 Torque and PE of coils in angular displacement: The system shown in Fig.4.22 was explained earlier in Sec.4.14. Now the coil carries N_2I_2 ampere-turns. Clockwise torque on the coil $T = N_2I_2A.B.Sin\theta$ follows a sine law with the angle. With rotational or circular motion, it is convention to reckon anticlockwise direction as positive, and so the torque is negative here, $T = -N_2I_2A.B.Sin\theta$. The situation is similar to that in Fig.4.21(d). The zero reference position of PE_M of the coil is AA', cumulative alignment position with the field.

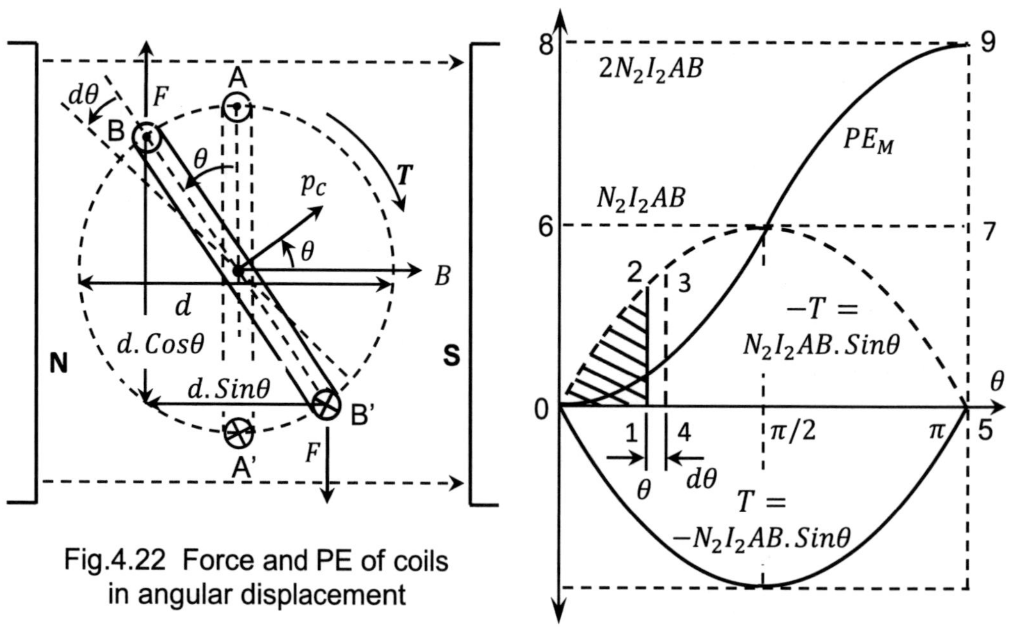

Fig.4.22 Force and PE of coils in angular displacement

As the coil rotates through an angle $+d\theta$, the PE_M of the coil increases by

$$dPE_M = (-T)(+d\theta) = -Td\theta = \text{Annular area 1234 and we write}$$

$$-T = \frac{+dPE_M}{+d\theta} \qquad \text{..............................} (4.36)$$

to interpret that *when the torque T is negative (clockwise), PE_M increases with increase in the angle θ* (in the anticlockwise direction). PE_M of the coil at its BB' position

$$PE_M = \int_0^\theta (-T)d\theta \text{ (shaded area 012) } = N_2I_2A.B\int_0^\theta Sin\theta.d\theta$$
$$= N_2I_2A.B(1 - Cos\theta) = p_c.B(1 - Cos\theta) \quad \text{.............. (4.37)}$$

The plot of PE_M shows that as the coil rotates from $\theta = 0$ (coil in cumulative alignment with the field) to $\theta = \pi$ (coil in differential alignment with the field), the PE_M increases from zero to $2N_2I_2A.B\ (= 2.p_c.B)$. The opposing flux of the field through the coil in its position AA' is $-A.B$, and that in its position BB' is $-A.B.Cos\theta$, an increase by $\Delta\phi_{12} = (-A.B.Cos\theta) - (-A.B) = A.B(1 - Cos\theta)$. Hence, Eqn.4.35, $PE_M = N_2I_2.\Delta\phi_{12}$ remains complied.

Ch. 5: Electromagnetic Induction

5.1 Introduction: The phenomenon of producing electricity from magnetism is referred to as Electromagnetic Induction. Whenever we come across this term, the person we recall first is the British book binder apprentice-turned-scientist Michael Faraday (1791-1867) who himself was discovered by the famous British scientist Sir Humphrey Davy (1778-1829). After the discovery of Electromagnetism, that is, electric current produces magnetism, by Oersted in 1820, Faraday, in 1821, started his research on the possibility of the reverse phenomenon, that is, whether magnetism can also produce electricity, which is known subsequently as Electromagnetic Induction. On August 29, 1831 he announced his discovery of electromagnetic induction, most simply, using only pictures and lines of force without any mathematics. Although the American physicist Joseph Henry (1797-1878), and also some other physicists, observed electromagnetic induction independently by the same period of time, Faraday became the officially recognised discoverer of the phenomenon as he published his results first. This brilliant discovery of Faraday brings electricity to our houses today and makes electricity commercially available for business and industrial purposes. It is learnt that once on asking Davy described Faraday as his greatest discovery. Faraday is crowned with the title 'Father of Electricity'.

5.2 Faraday's Law of Electromagnetic Induction: This law states that *whenever magnetic flux linking (threading or passing through) an electric circuit (coil or winding) changes, or a conductor cuts (or is cut by) magnetic flux, an emf is induced in the circuit or in the conductor.* Faraday also showed that the magnitude of the induced emf is *proportional* to the (time-) rate of change of the flux linkage or of the flux cut. It will become evident in Sec.5.4 that the constant of proportionality is unity, and hence, the magnitude of the induced emf is *equal to* the rate of change of the flux linkage or of the flux cut, as the case may be.

Principles, methods and processes of EM induction have been grouped in Fig.5.1. All forms of EM induction employ flux linkage principle in common, and these may occur by both static and dynamic methods. If occurs by static method, this leads to the generation of *statically induced emf*, wherein an emf is induced in a static coil by a varying flux through the coil. If this varying flux results from a varying current through the same coil, *self induction* occurs as a process, and if it results from a varying current through some neighbouring coil, *mutual induction* occurs. Hence, statically induced emf may occur by self or mutual induction. EM induction occurring by dynamic methods employs flux cut principle, which in fact is not different from the flux linkage principle but is only a different view of the latter, so can always be explained in terms of the latter (Sec.5.7). *Dynamically induced emf* is generated in a conductor or coil as it cuts (or is cut by) magnetic flux. Dynamically induced emf occurs by mutual induction only, and self induction is not possible by dynamic means. It follows that self induction involves only one system (one coil), whereas mutual induction involves two systems (two coils or one coil and one magnet).

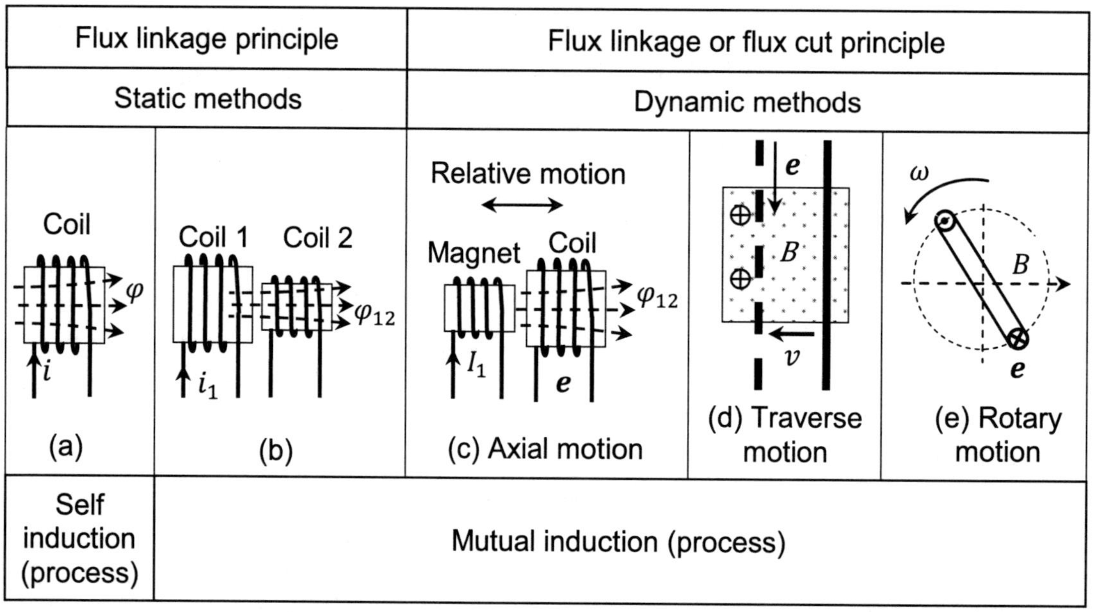

Fig. 5.1 Grouping of electromagnetic induction

5.3 Lenz's Law: The direction of the induced emf was formulated by the Russian physicist Heinrich Friedrich Emil Lenz (1804-1865) in 1833. This is known as Lenz's Law which states that *any electromagnetically induced emf always tends to set up a current in such a direction that opposes the very cause to which the emf is due*. This will also be explained in Sec.5.4.

5.4 Self Induction and Self Inductance: *Self induction is the process of inducing an EM emf in an electric circuit caused by a change in its flux linkage due to some changing current in the same circuit.* In Fig.5.2, suppose the current through the N-turn coil at any instant of time t is i, and this causes a flux φ through the coil. If the current increases by di in time dt, the flux increases by $d\varphi$. As was discussed in Sec.4.7.1, the SE_M in the magnetic field associated with the coil increases by $dSE_M = Ni.d\varphi$. This energy must be supplied by the DC source (battery). Ohmic heating is also there in the resistance R in the circuit. Energy equation during time dt,

Battery output = Increase in SE_M + Ohmic heat in R

or, $E.i.dt = Ni.d\varphi + i^2R.dt$

or, $E = N.d\varphi/dt + iR$

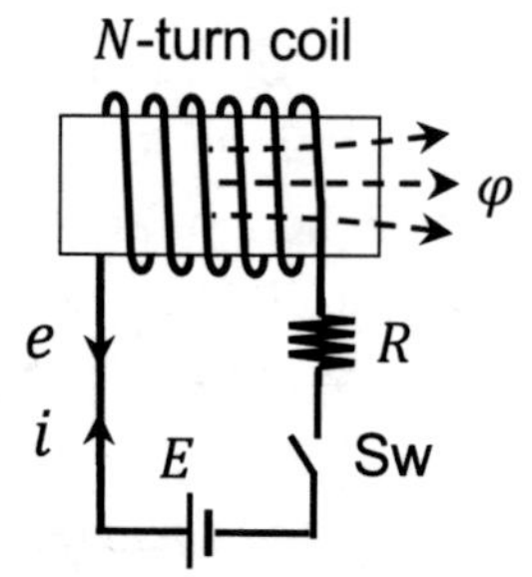

Fig.5.2 Self induction

This being a voltage equation, the quantity $N.d\varphi/dt$ must be some voltage or emf

$e = N.d\varphi/dt$ = Rate of increase in flux linkage

generated in the coil, which is Faraday's self induced emf. Further, to satisfy the voltage identity $E - e = iR$, the emf e must act in opposition to E or i, to oppose i to increase, to which the emf e is due. This is Lenz's law.

If i decreases, e acts along i, to oppose i to decrease. This opposing feature of Lenz's law introduces a negative sign in Faraday's expression of EM emf, as

$$e = -Nd\varphi/dt$$

Faraday's law, in conjunction with Lenz's law can be expressed as

e in the direction of $i =$ − Rate of increase of flux linkage

= Rate of decrease of flux linkage

It is thus seen that both Faraday's law and Lenz's law follow from the Law of Conservation of Energy, and this applies to all cases of electromagnetic induction.

If S be the reluctance of the magnetic circuit of the coil, $\varphi = \frac{Ni}{S}$, or $d\varphi = \frac{N}{S}di$,

and as such, $e = -N\dfrac{d\varphi}{dt} = -\dfrac{N^2}{S}\dfrac{di}{dt}$

The quantity $\dfrac{N^2}{S} = \dfrac{N}{i}\dfrac{Ni}{S} = \dfrac{N}{i}\varphi = \dfrac{N\varphi}{i}$ (dimensionally $\mu.l$) that represents 'flux linkage per unit current (ampere)' through the same coil is defined as its *self inductance* L. Hence we have

$$L = \frac{N\varphi}{i} = \frac{N^2}{S} \qquad \text{.........................} \ (5.1)$$

$$\text{and, } e = -N\frac{d\varphi}{dt} = -L\frac{di}{dt} \qquad \text{.........................} \ (5.2)$$

Faraday discovered self inductance in 1834, after three years he discovered electromagnetic induction in the form of mutual induction in 1831.

Inductance is a property of a coil or electric circuit, like its resistance. It, however, varies with permeability (Sec.6.3). Inductance opposes any change in the current through the circuit. It is analogous to the inertia of a mass in Mechanics that opposes any change in the velocity of the mass. Eqn.5.1 reveals that the unit of inductance is weber per ampere (Wb/A). This has been named henry (H) after Joseph Henry.

Substituting $N\phi = LI$ in Eqn.4.15 (Sec.4.7.1), we get the energy stored in electromagnetic field in an alternative form

$$SE_M = \frac{1}{2}NI\phi = \frac{1}{2}LI^2 \qquad \text{............................} \ (5.3)$$

This is the I- concept in Electromagnetic Induction, as was mentioned in Sec.1.14.

Inductance in Electromagnetic Induction is analogous to capacitance in Electrostatics.

As $\phi(= m)$ is analogous to Q, and

mmf NI (dimensionally I) is analogous to V,

$L = N\phi/I$ (dimensionally ϕ/I) is analogous to $C = Q/V$

Further, like self capacitance and mutual capacitance in Electrostatics, here we have the analogous quantities self inductance and mutual inductance respectively.

5.4.1 Switching ON an inductive circuit: When the power to an inductive circuit (coil) is switched on, Fig.5.2, the current grows from zero to the final value $I = E/R$. As the inductance of the coil impedes this growth, the current cannot grow instantly, but is delayed, as shown by the i-curve in Fig.5.3(a).

The situation is exactly similar to the growth of voltage across a capacitor, Sec.2.11, and present derivations follow similar lines.

$$E - iR = L\frac{di}{dt} \quad \text{or, } \frac{E}{R} - i = \frac{L}{R}\frac{di}{dt}$$

$$\text{or, } I - i = \frac{L}{R}\frac{di}{dt} \quad \text{or, } \frac{R}{L}dt = \frac{di}{I-i}$$

Integrating both sides, $\frac{R}{L}t = -log_e(I - i) + C$,

where the mathematical constant $e = 2.7183$ is the base of natural logarithm (not the induced emf), and C is the constant of integration.

At the instant $t = 0,\ i = 0$, so $C = log_e I$. Hence,

$$\frac{R}{L}t = -log_e(I - i) + log_e I = log_e \frac{I}{(I-i)}$$

$$\text{or, } \frac{I-i}{I} = e^{-Rt/L} \quad \text{or, } i = I(1 - e^{-Rt/L})$$

$$\text{or, } i = I(1 - e^{-t/T}) \qquad \ldots\ldots\ldots\ldots\ldots\ldots\ldots (5.4)$$

where $T = L/R$ (dimensionally time) is called the *time constant* of the circuit. This exponential relation is due to the German physician, physicist and philosopher Helmholtz (1821-1894), so is often referred to as Helmholtz equation.

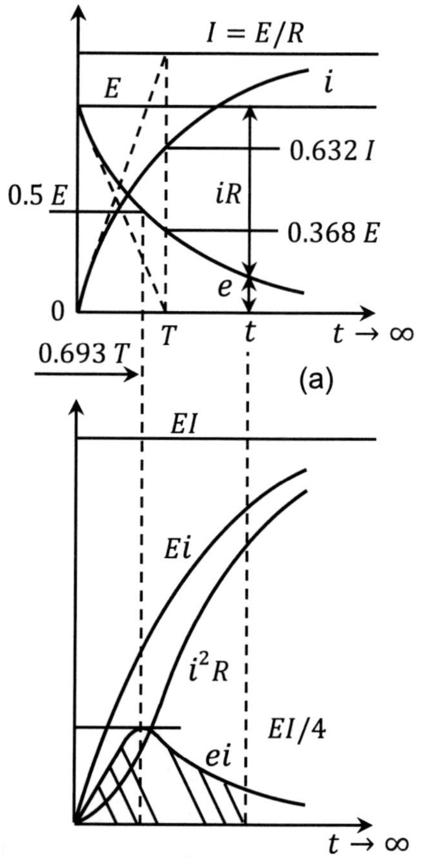

Fig.5.3 Switching ON an inductive circuit

At the instant of switching $(t = 0)$, $i = 0,\ iR = 0,\ e = E$. As the current grows, iR increases and e falls off. At the instant t

$$\begin{aligned} e &= E - iR \\ &= E - I(1 - e^{-t/T})R \\ &= Ee^{-t/T} \qquad \ldots\ldots\ldots\ldots\ldots\ldots\ldots\ldots (5.5) \end{aligned}$$

Thus, e first rises to E and then decays exponentially, as shown in Fig.5.3(a).

The time constant T has the same significance as in the corresponding electric case. We had

$$\frac{R}{L}dt = \frac{di}{I-i}$$

Rearranging this, $\frac{di}{dt} = \frac{I-i}{L/R} = \frac{I-i}{T}$. Thus, referring to Fig.5.4, at any instant t during growth of current in an inductive circuit if the current i continues to grow with the instantaneous rate di/dt, the current will attain its final value I in next time constant period T.

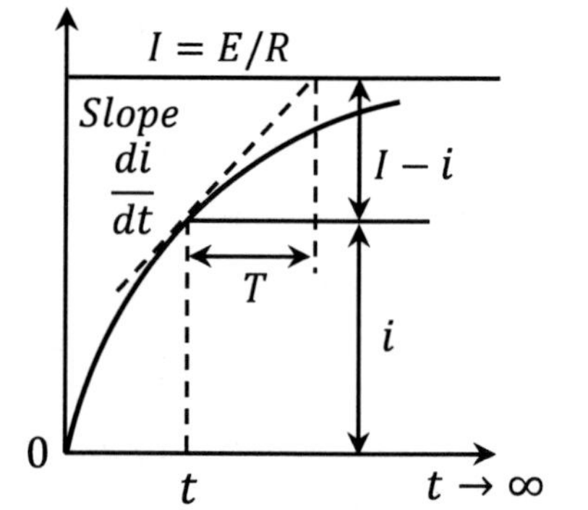

Fig. 5.4 Time constant of inductive circuit

The current at the instant T $(t = T)$,

$$i = I(1 - e^{-1}) = I(1 - 0.368) = 0.632\,I$$

In words, the current grows to 63.2% of its final value during initial time constant period, as indicated in Fig.5.3(a). The current grows to 99.33% of its final value during initial $5T$ time period $(t = 5T)$ and to 99.99% of its final value during initial $10T$ time period.

Similarly, the induced emf at the instant $t = T$

$$e = Ee^{-1} = 0.368\,E$$

That is, the induced emf falls to 36.8% of its initial value during initial time constant period, to 0.673% during initial $5T$ time period, and to 0.004% during initial $10T$ time period.

The curves in Fig.5.3(b) explains the power balance during the switching ON period of the circuit. As was mentioned in Sec.5.4, out of the total power Ei supplied by the source (battery) at any instant t, a portion is consumed as i^2R heating, and the rest ei contributes to the increase in the SE_M in the magnetic field. That is, $Ei = ei + i^2R$. Peculiar trend of the ei curve is similar to the electric case, Sec.2.11. As $e + iR = E$ is constant, the product $e.iR$ exhibits a maximum value, or the product $e.i$ exhibits a maximum value (R being constant), when $e = iR = 0.5E$. Hence the condition for this maximum value to occur

$$e = Ee^{-t/T} = 0.5E, \quad \text{or, } e^{-t/T} = 0.5 \quad \text{or, } e^{t/T} = 2$$

$$\text{or, } t/T = log_e 2 = log_{10}2/log_{10}e = 0.3010/0.4343 = 0.693$$

$$\text{or } t = 0.693\,T$$

Putting $e^{-t/T} = 0.5$ in the equation $ei = Ee^{-t/T}.\, I(1 - e^{-t/T})$, we get the maximum value of ei as

$$(ei)_{max} = \frac{EI}{4} \qquad \text{.. (5.6)}$$

and this occurs at

$$t = 0.693\,T \qquad \text{.. (5.7)}$$

as shown in Fig.5.3(b). The shaded area below the ei curve

$$\int_0^t eidt = L\int_0^t \frac{di}{dt} idt = L\int_0^i i\,di = \frac{1}{2}Li^2$$

gives the SE_M in the magnetic field upto the period the current has grown to the value i. As this is proportional to i^2, the SE_M energy curve (not shown in the figure) is the same as the i^2R power curve drawn to some different scale. When the current grows to its final value I,

$$SE_M = \frac{1}{2}LI^2$$

as was given by Eqn.5.3. Theoretically, same as in the electric case, it takes infinite time for the above growths and decays to their final values, but in practice these happen in seconds or even in a fraction of second.

5.4.2 Switching OFF an inductive circuit: If an inductive circuit like one shown in Fig.5.2 is switched off, the current will drop from its steady value I to zero, and the resulting induced emf e will add to the battery voltage. The quantum of the arc in the switch, which is proportional to the circuit voltage, will therefore enhance. In fact, the arc, which in effect is the circuit current, will not be extinguished till the stored energy $LI^2/2$ is entirely exhausted as heat in the arc and the circuit resistance. If the amount of stored energy is large, the arc will take the form of a flash. Hence it may be dangerous to switch off a large electromagnet as such. As one remedy, a resistance can be inserted across the electromagnet to convert the stored energy into heat. But this causes continuous waste of energy in this resistance so long the electromagnet remains on. Another remedy, as usually employed in the field circuit of DC machines, is shown in Fig.5.5(a).

A field discharge resistance (FDR) of resistance value r is connected across the field circuit through the field discharge switch (FDS). The FDS is a 2-way switch, ON and OFF, and it is so designed that switching from ON position to OFF position occurs with close-circuit transition. Thus, when the switch is turned from ON to OFF, contact 2 makes before contact 1 breaks, as indicated in the figure, thereby using the FDR only during the switching off period of the field circuit. The switching off operation, therefore, comprises of two sequences.

(a) Switching from the ON position (making contact 1, connecting only the field circuit to battery) to the momentary intermediate position (making both the contacts 1 and 2, thereby adding the FDR circuit to battery). This is *ready to switch off sequence*.

(b) Switching from the momentary position to OFF position (contact 2 remains made and contact 1 opens, disconnecting both the circuits from the battery). This is actual *switch off sequence*.

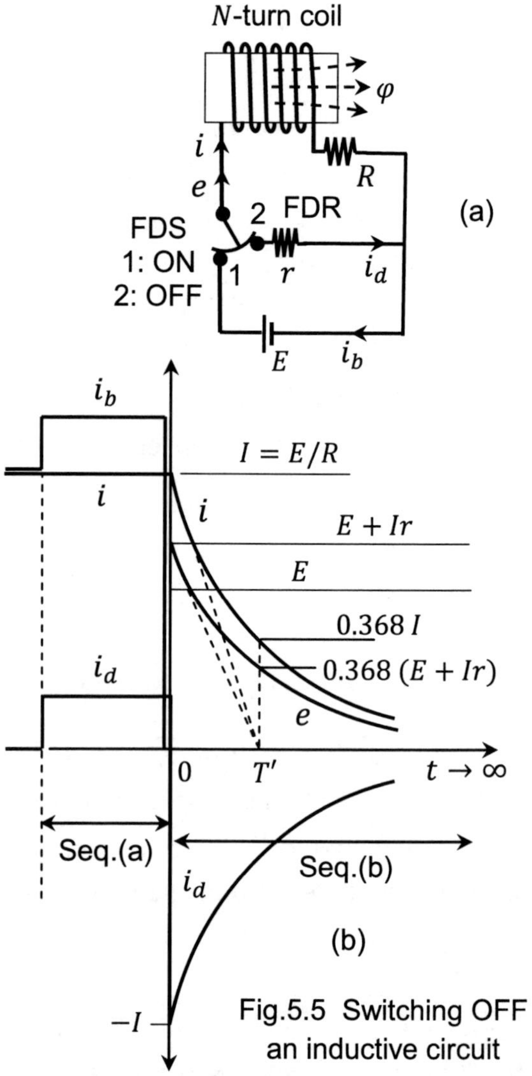

Fig.5.5 Switching OFF an inductive circuit

These sequences are explained by the graphs in Fig.5.5(b). Let,

i = Field current

i_d = Current through the FDR in the direction shown, and

i_b = Battery current

In Seq.(a), i remains at its steady value $I = E/R$, and i_b increases by the steady current $I_d = E/r$ through the FDR, from I to $(I + I_d)$. In Seq.(b), I_b instantaneously drops to zero, but the induced emf in the field winding will not allow i to do so. Hence the FDR will carry a reverse current $i_d = -i$. Reckoning from zero instant of time $(t = 0)$ at the start of Seq.(b), voltage equation for the field-FDR loop at any instant of time t is

$$e = L\frac{di}{dt} = i(R + r) = iR'$$

Where di is decrease in the field-FDR loop current, and $R' = R + r$. Hence,

$$\frac{R'}{L}dt = \frac{di}{i}$$

As t increases by dt, i decreases by di, and this necessitates a minus sign for integration, same as in earlier occasions like Sec.1.8.2(b).

Thus we have $\frac{R'}{L}dt = -\frac{di}{i}$. Integrating both sides,

$$\frac{R'}{L}t = -log_e i + C,$$

where C is the constant of integration. At the instant $t = 0$, $i = I$, so $C = log_e I$.

Therefore, $\frac{R'}{L}t = -log_e i + log_e I = log_e(I/i)$

or, $\frac{I}{i} = e^{R't/L}$ or, $\frac{i}{I} = e^{-R't/L}$

or, $i = Ie^{-R't/L} = Ie^{-t/T'}$ .... (5.8)

$T' = L/R' = L/(R + r)$ is the time constant of the switch off circuit (which was $T = L/R$ for the switch on circuit). This means, the current decays exponentially with a faster rate compared to its switch on growth rate. Further,

$e = iR' = I(R + r)e^{-t/T'}$

$= (E + Ir)e^{-t/T'}$ .... (5.9)

This shows that at the start of Seq.(b), a transient voltage spike $e = (E + Ir)$, which is higher than E, strikes the field-FDR loop, and it then decays away exponentially at a faster rate compared to its switch on decay rate.

5.5 Mutual Induction: *Mutual induction can be defined as the process of inducing an electromagnetic emf in an electric circuit caused by a change in its flux linkage due to some neighbouring system.* Hence mutual induction originates from the proximity or closeness between two neighbouring systems. It may occur both by static and dynamic methods, and may utilise flux linkage or flux cut principle.

5.6 Mutual Induction by Static Method – Mutual Inductance: Fig.5.6 explains mutual induction by static method. A changing current i_1 in coil 1 not only changes the flux linkage of the same coil, but also of the neighbouring coil 2. Hence statically induced emfs are generated in both the coils, by self induction in coil 1, and by mutual induction in coil 2. If N_1 and N_2 be the number of turns on coil 1 and coil 2 respectively, and S be the reluctance of the common magnetic circuit of the two coils, the changing mutual flux produced by coil 1 that links all the turns of coil 2

$$\varphi_{12} = N_1 i_1 / S$$

This changing mutual flux φ_{12} generates in coil 2 a *mutually induced emf*

$$e_2 = \frac{d(N_2\varphi_{12})}{dt} = N_2\frac{d\varphi_{12}}{dt} = \frac{N_1 N_2}{S}\frac{di_1}{dt}$$

in the direction given by Lenz's Law. The quantity

$$\frac{N_1 N_2}{S} = \frac{N_2}{i_1}\frac{N_1 i_1}{S} = \frac{N_2}{i_1}\varphi_{12} = \frac{N_2\varphi_{12}}{i_1}$$

that represents the 'flux linkage of coil 2 per unit current (ampere) through coil 1' is defined as the mutual inductance M_{12} between the coils when coil 1 is primary. Hence we have

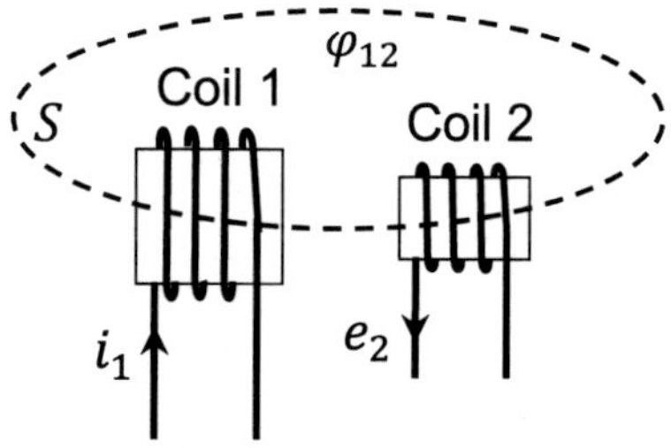

Fig.5.6 Mutual induction by static method

$$M_{12} = \frac{N_1 N_2}{S} \qquad \text{............................ (5.10)}$$

$$e_2 = N_2\frac{d\varphi_{12}}{dt} = M_{12}\frac{di_1}{dt} \qquad \text{............................ (5.11)}$$

Similarly, we may derive the mutual inductance between the coils when coil 2 is primary, as

$$M_{21} = \frac{N_1 N_2}{S}$$ (This also follows from the symmetry of Eqn.5.10).

Hence, $M_{12} = M_{21} = M$ .... (5.12)

This identity can also be obtained from our preparatory concept in Sec.4.17.1. We had

$$N_1 I_1 \phi_{21} = N_2 I_2 \phi_{12} \text{ (Eqn. 4.29) or, } \frac{N_2\phi_{12}}{I_1} = \frac{N_1\phi_{21}}{I_2}$$

Thus, $M_{12} = \frac{N_2\varphi_{12}}{i_1} = \frac{N_1\varphi_{21}}{i_2} = M_{21}$

This reciprocity rule of mutual inductance implies that the *mutual inductance between two circuits remains the same no matter which circuit is used as primary.*

5.6.1 Relation between self inductance and mutual inductance – Coupling Coefficient: When two current carrying coils neighbour, and in the general case, they have leakage fluxes, we utilise our preparatory concept developed in Sec.4.17.1 to write the following relations. Self inductance of coil 1,

$$L_1 = \frac{N_1^2}{S_1} = \frac{N_1^2}{S_{1L}} + \frac{N_1^2}{S} = L_{1L} + L_{1U}$$

L_{1L} is *leakage inductance* of coil 1. We may name L_{1U} as *useful inductance* of coil 1, which, as seen below, is responsible for its mutual inductance to coil 2.

Similarly, self inductance of coil 2

$$L_2 = \frac{{N_2}^2}{S_2} = \frac{{N_2}^2}{S_{2L}} + \frac{{N_2}^2}{S} = L_{2L} + L_{2U}$$

Also, mutual inductance between the coils,

$$M = \frac{N_1 N_2}{S} = \sqrt{L_{1U} L_{2U}}$$

Hence, two mutually coupled coils are represented by the equivalent circuit shown in Fig.5.7.

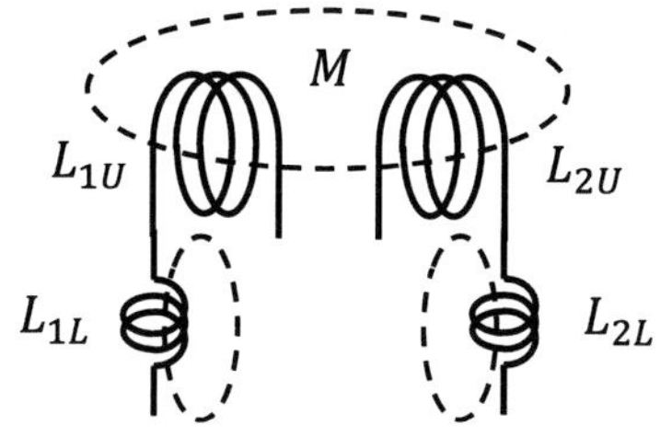

Fig.5.7 Equivalent circuit of coupled coils with leakage

Above relations may be rewritten as

$$L_1 = \frac{{N_1}^2}{S_{1L}} + \frac{{N_1}^2}{S} = \frac{{N_1}^2(S+S_{1L})}{S.S_{1L}} = L_{1U}\frac{(S+S_{1L})}{S_{1L}}$$

$$L_2 = \frac{{N_2}^2}{S_{2L}} + \frac{{N_2}^2}{S} = \frac{{N_2}^2(S+S_{2L})}{S.S_{2L}} = L_{2U}\frac{(S+S_{2L})}{S_{2L}}$$

Multiplying these equations,

$$L_1.L_2 = L_{1U}L_{2U}\frac{(S+S_{1L})}{S_{1L}}\frac{(S+S_{2L})}{S_{2L}} = M^2\frac{(S+S_{1L})(S+S_{2L})}{S_{1L}.S_{2L}}$$

$$\text{or, } M^2 = L_1 L_2 \left(\frac{S_{1L}}{S+S_{1L}}\right)\left(\frac{S_{2L}}{S+S_{2L}}\right)$$

It can be seen that the fraction of the coil 1 flux that links coil 2, $k_1 = \frac{\phi_{12}}{\phi_1} = \frac{S_{1L}}{S+S_{1L}}$

And the fraction of the coil 2 flux that links coil 1, $k_2 = \frac{\phi_{21}}{\phi_2} = \frac{S_{2L}}{S+S_{2L}}$

Hence, $M = \sqrt{k_1.k_2}\sqrt{L_1 L_2} = k\sqrt{L_1 L_2}$ .... (5.13)

relates the self inductances of two neighbouring coils with their mutual inductance. The fraction k, whose value lies between 0 to 1 ($0 < k < 1$), is called the *coefficient of coupling* or *coupling coefficient* of the coils. This term is mostly used in electronic circuits. If this factor k is less than 0.5, the coils are said to be *loosely coupled*, and if it is more than 0.5, *tightly coupled*.

5.6.2 Inductances in series:

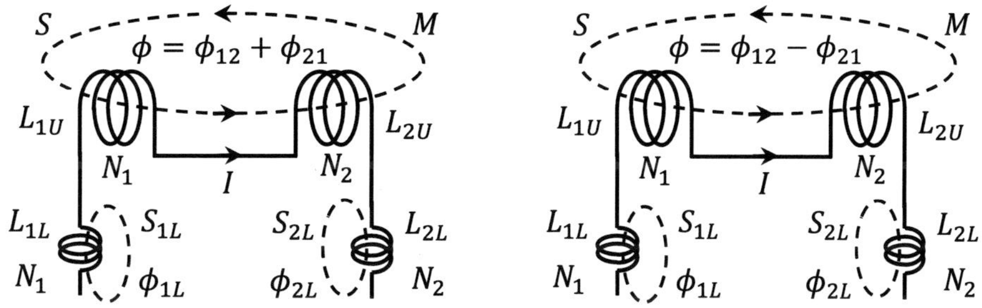

(a) Cumulatively coupled $L_C = L_1 + L_2 + 2M$

(b) Differentially coupled $L_D = L_1 + L_2 - 2M$

Fig.5.8 Mutually coupled coils in series

Fig.5.8(a) shows the equivalent circuit of two mutually coupled coils (inductances) connected in series. The coils are so connected that they are in cumulative coupling. For a current I through the coils, the fluxes are as shown. The flux through the common magnetic circuit of reluctance S is $\phi = \phi_{12} + \phi_{21}$ (Sec.4.17.1). The equivalent inductance of the combination can be evaluated as the total flux linkage per unit current.

$$L_C = \frac{N_1\phi_{1L}}{I} + \frac{N_1\phi}{I} + \frac{N_2\phi}{I} + \frac{N_2\phi_{2L}}{I}$$

$$= \frac{N_1\phi_{1L}}{I} + \frac{N_1\phi_{12}}{I} + \frac{N_1\phi_{21}}{I} + \frac{N_2\phi_{12}}{I} + \frac{N_2\phi_{21}}{I} + \frac{N_2\phi_{2L}}{I}$$

$$= \frac{N_1^2}{S_{1L}} + \frac{N_1^2}{S} + \frac{N_1N_2}{S} + \frac{N_1N_2}{S} + \frac{N_2^2}{S} + \frac{N_2^2}{S_{2L}}$$

$$= \left[\frac{N_1^2}{S_{1L}} + \frac{N_1^2}{S}\right] + \left[\frac{N_2^2}{S_{2L}} + \frac{N_2^2}{S}\right] + 2\left[\frac{N_1N_2}{S}\right]$$

$$L_C = [L_{1L} + L_{1U}] + [L_{2L} + L_{2U}] + 2M$$

$$= L_1 + L_2 + 2M \qquad \text{.................................. (A)}$$

When the coils are differentially coupled, Fig.5.8(b), we have $\phi = \phi_{12} - \phi_{21}$, and following similar derivation, the equivalent inductance of the two coils

$$L_D = L_1 + L_2 - 2M \qquad \text{.................................. (B)}$$

Combining these two results, Eqns. A and B, the equivalent inductance of two mutually coupled coils,

$$L_S = L_1 + L_2 \pm 2M \qquad \text{.............................. (5.14)}$$

Positive sign applies for cumulative coupling and negative sign for differential coupling. We prove below L_S is always positive, or $(L_1 + L_2) > 2M$. From previous Sec.5.6.1,

$$L_1 = L_{1L} + L_{1U} = L_{1L} + \frac{N_1^2}{S} = L_{1L} + \frac{N_1N_2}{S}\cdot\frac{N_1}{N_2} = L_{1L} + M\frac{N_1}{N_2}$$

And similarly, $L_2 = L_{2L} + M\frac{N_2}{N_1}$.

Adding these, $L_1 + L_2 = L_{1L} + L_{2L} + M\left[\frac{N_1}{N_2} + \frac{N_2}{N_1}\right]$

Now, N_1/N_2 and N_2/N_1 are positive numbers and reciprocal of one another. It is mathematics that the sum of a positive number and its reciprocal is ≥ 2; if the number is 1 (here $N_1 = N_2$), the sum is 2 and as the number increases, the sum increases above 2. Hence,

$$L_1 + L_2 \geq L_{1L} + L_{2L} + 2M$$

Thus, even in the extreme theoretical case of zero magnetic leakage between the coils,

$$L_1 + L_2 \geq 2M.$$

The sum $L_1 + L_2$ will equal $2M$ if the coils be in exact physical coincidence, an arrangement practically impossible to achieve.

When n number of coils are in series, and are not coupled ($M = 0$), their equivalent inductance

$$L_S = L_1 + L_2 + L_3 + \ldots L_n \qquad \text{.................................. (5.15)}$$

If, for example, three coils 1, 2, 3 of self inductances L_1 L_2 L_3 are in series, coils 1, 2 are cumulatively coupled with mutual inductance M_{12}, and coils 2, 3 are differentially coupled with mutual inductance M_{23}, it is logical to conclude from Eqn.5.14 that the equivalent inductance of the combination becomes

$$L_S = L_1 + L_2 + L_3 + 2M_{12} - 2M_{23}$$

5.6.3 Inductances in parallel: Fig.5.9 shows two cumulatively coupled coils connected in parallel. Let L be the equivalent inductance of the combination. A varying current i fed to the combination is so divided into i_1 and i_2 through the two branches that, neglecting circuit resistances, the induced emf in each branch equals that in L.

$$L\frac{di}{dt} = L_1\frac{di_1}{dt} + M\frac{di_2}{dt} = L_2\frac{di_2}{dt} + M\frac{di_1}{dt}$$

or, $Li = L_1 i_1 + M i_2 = L_2 i_2 + M i_1$ .... (A)

or, $L_1 i_1 = Li - Mi_2$ and $L_2 i_2 = Li - Mi_1$

Adding, $L_1 i_1 + L_2 i_2 = 2Li - M(i_1 + i_2) = 2Li - Mi$

$= (2L - M)i$ .... (B)

Also from Eqn.A, $i_1 = \dfrac{Li - Mi_2}{L_1}$ and $i_2 = \dfrac{Li - Mi_1}{L_2}$

Adding, $i_1 + i_2 = i = \dfrac{LL_2 i - ML_2 i_2 + LL_1 i - ML_1 i_1}{L_1 L_2}$

or, $L_1 L_2 i = Li(L_1 + L_2) - M(L_1 i_1 + L_2 i_2)$

$= Li(L_1 + L_2) - M(2L - M)i$ (by Eqn.B)

or, $L_1 L_2 = L(L_1 + L_2) - M(2L - M)$

$= L(L_1 + L_2 - 2M) + M^2$

or, $L = \dfrac{L_1 L_2 - M^2}{L_1 + L_2 - 2M}$

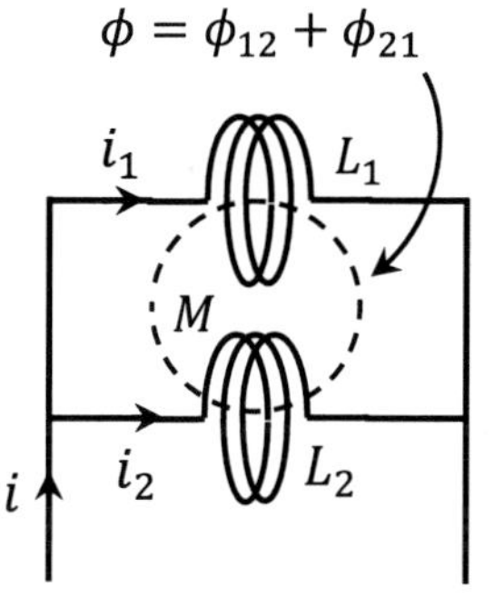

Fig.5.9 Cumulatively coupled coils in parallel

If the coils are differentially coupled, similar derivation will result $L = \dfrac{L_1 L_2 - M^2}{L_1 + L_2 + 2M}$

Combining into a single equation, the equivalent inductance of the parallel combination

$$L_P = \frac{L_1 L_2 - M^2}{L_1 + L_2 \mp 2M}$$

It is convenient to write in the form

$$\frac{1}{L_P} = \frac{L_1 + L_2 \mp 2M}{L_1 L_2 - M^2} \quad \text{.................................. (5.16)}$$

Negative sign applies for cumulative coupling and positive sign for differential coupling. Hence, when n number of coils are in parallel, and they are not coupled ($M = 0$), their equivalent inductance is given by

$$\frac{1}{L_P} = \frac{1}{L_1} + \frac{1}{L_1} + \frac{1}{L_3} + \cdots \frac{1}{L_n} \quad \text{.................................. (5.17)}$$

Series parallel combination of inductances thus obeys the same laws as for resistances.

If, for example, three coils 1, 2, 3 of self inductances L_1, L_2, L_3 are in parallel, coils 1, 2 are cumulatively coupled with mutual inductance M_{12}, and coils 2, 3 are differentially coupled with mutual inductance M_{23}, the pattern of the Eqns.5.16 and 5.17 suggests that the equivalent inductance of the combination will be given by

$$\frac{1}{L_P} = \frac{L_1L_2+L_2L_3+L_3L_1-2M_{12}+2M_{23}}{L_1L_2L_3-M_{12}{}^2-M_{23}{}^2}$$

5.6.4 The Transformer Principle: Let us close the coil 2 circuit of Fig.5.6, as in Fig.5.10. Situations *without* and *with* magnetic leakage of the coils need separate consideration.

(a) Without magnetic leakage: Fig.5.10(a) explains this situation. On closing the switch, i_0 grows in circuit 1, and the common flux φ_{12} that i_0 produces, induces the emfs e_1 and e_2 in the two coils. The variations of the quantities have also been shown in the figure. As circuit 2 is closed with its resistance R_2, current $i_2 = e_2/R_2$ starts with a peak, and then decays to zero with e_2. Heat energy consumed by R_2 in the process has to be supplied by the only source in the system, the cell or battery in circuit 1. This requires an additional current i_1' drawn from the battery in circuit 1, and it will be seen below that energy balance requirement is fulfilled with the magnitude of the current i_1' given by $N_1i_1' = N_2i_2$, where N_1 and N_2 are the number of turns on coil 1 and 2 respectively. These mmfs can be seen to act in opposition on the common magnetic circuit, and so refrain from disturbing φ_{12} growing proportionally with i_0 at any instant of time. However, the additional voltage drop $i_1'R_1$ delays the growth of i_0 and φ_{12} compared to when circuit 2 was open. Total current in circuit 1 is thus given by $i_1 = i_0 + i_1'$. Voltage equation of circuit 1 is

$$E_1 = i_1R_1 + e_1 \text{ or, } i_1 = \frac{E_1-e_1}{R_1} = I_1 - \frac{e_1}{R_1}$$

$I_1 = E_1/R_1$ being the circuit 1 final steady current. This shows that i_1 is lower than I_1 and, therefore, i_1' is still lower. Hence i_1' and i_1 have the shapes as shown. As e_1 decays, i_1 grows to its final value I_1. Analytical treatment of the topic follows.

We have, $e_1 = N_1\frac{d\varphi_{12}}{dt}$, $e_2 = N_2\frac{d\varphi_{12}}{dt}$ and $N_1i_1' = N_2i_2$. Hence we have

$$\frac{e_1}{e_2} = \frac{N_1}{N_2} = \frac{i_2}{i_1'}$$

$$\text{and, } e_1i_1' = e_2i_2 \qquad \text{.......................... (5.18)}$$

At any instant of time t during the flux grows from zero to φ_{12}, voltage equation of circuit 2 is $e_2 = i_2R_2$.

Heat energy consumed by R_2 during time dt

$$e_2i_2dt = i_2^2R_2dt \qquad \text{............................ (A)}$$

Similarly, voltage equation of circuit 1 being

$$E_1 = i_1R_1 + e_1$$

energy supplied by the battery during time dt

$$E_1 i_1 dt = i_1^2 R_1 dt + e_1 i_1 dt$$
$$= i_1^2 R_1 dt + e_1(i_0 + i_1^{'})dt$$
$$= i_1^2 R_1 dt + e_1 i_0 dt + e_1 i_1^{'} dt$$
$$= i_1^2 R_1 dt + e_1 i_0 dt + e_2 i_2 dt \quad \text{(using Eqn.5.18)}$$

Of the right hand terms of this equation, the first one is heat energy consumed by R_1. The second term is increase in energy stored in coil 1, dSE_M, integral of which gives

$$\int_{t=0}^{t=\infty} dSE_M = \int_0^{\infty} e_1 i_0 dt = \int_0^{\infty} L_1 \frac{di_0}{dt} i_0 dt = L_1 \int_0^{I_1} i_0 . di_0 = \frac{1}{2} L_1 I_1^{\,2}$$

The third term is important. This, by Eqn.A, gives the heat energy consumed in circuit 2, and this has been transferred from circuit 1. As transformers operate on this principle of transferring electrical energy from one circuit (primary) to its coupled circuit (secondary) by mutual induction, this is commonly referred to as *transformer principle* or *transformer action*.

(b) With magnetic leakage: Fig.5.10(b) explains this situation. Leakage inductances of the coils L_{1L} and L_{2L}, which were ignored in situation (a) above, are now considered and added in the circuits. As such, in circuit 1 we have in this case,

$$L_1 = L_{1U} + L_{1L}$$

e_1 (Total induced emf in L_1) $= e_{1U}$ (Induced emf in L_{1U}) $+ e_{1L}$ (Induced emf in L_{1L})

As φ_{12} grows with i_0, this induces e_{1U} in coil 1 and e_2 in coil 2, in the same way as in the case (a) above. Here we have, $e_{1U} = N_1 \frac{d\varphi_{12}}{dt}$, $e_2 = N_2 \frac{d\varphi_{12}}{dt}$ and $N_1 i_1^{'} = N_2 i_2$. Hence

we have $\quad \frac{e_{1U}}{e_2} = \frac{N_1}{N_2} = \frac{i_2}{i_1^{'}}$ and, $e_{1U} i_1^{'} = e_2 i_2$ .... (5.19)

As circuit 2 now contains the inductance L_{2L}, i_2 cannot grow instantaneously; it grows slowly with time to some peak i_{2m} , of lower magnitude compared to that in case (a) as shown by the dotted curve, and then decays to zero with e_2. Similar variation also happens with the circuit 1 current $i_1^{'}$ complying to the identity $N_1 i_1^{'} = N_2 i_2$, and hence the variation of circuit 1 total current $i_1 = i_0 + i_1^{'}$ has the shape as shown.

Here we have the following analysis. Voltage equation of circuit 2 is $e_2 = i_2 R_2 + e_{2L}$, and the energy equation of circuit 2 during time dt is

$$e_2 i_2 dt = i_2^2 R_2 dt + e_{2L} i_2 dt \quad \text{............................. (A)}$$

First term on the right hand side is heat in R_2. The second term requires an elaboration.

$$e_{2L} i_2 dt = L_{2L} \frac{di_2}{dt} i_2 dt = L_{2L} i_2 di_2 .$$

As i_2 varies from zero to zero, the integral $L_{2L} \int_0^0 i_2 . di_2$ is indeed zero. During i_2 grows to its peak i_{2m}, the inductance L_{2L} stores $SE_M = \frac{1}{2} L_{2L} i_{2m}^{\,2}$, and during decaying of i_2 subsequently, L_{2L} releases this stored energy which is consumed in heating R_2.

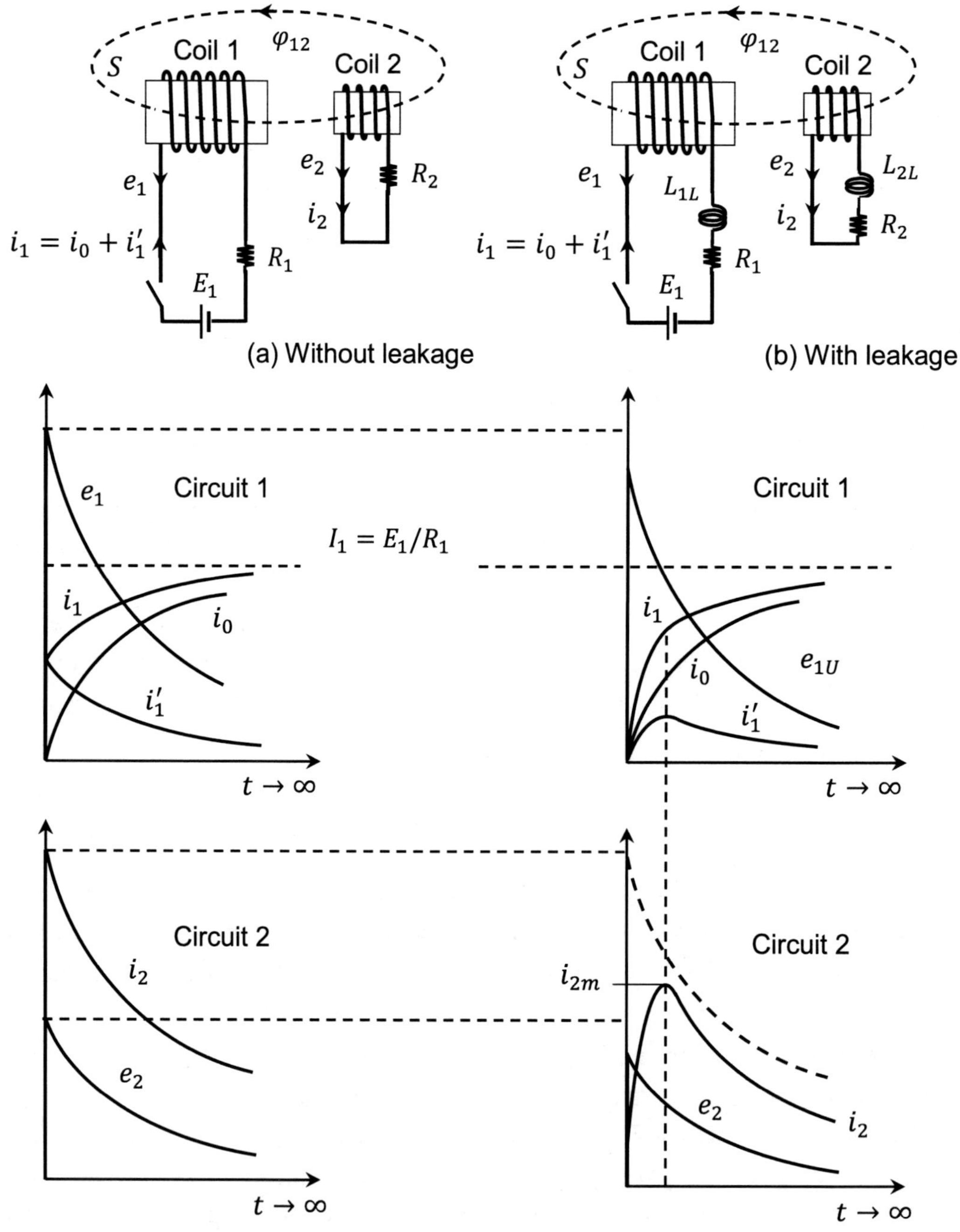

Fig.5.10 Transformer Principle explained

Voltage equation of circuit 1 is

$$E_1 = i_1R_1 + e_1 = i_1R_1 + e_{1U} + e_{1L}$$

and the energy supplied by the battery during time dt

$$E_1 i_1 dt = i_1^2 R_1 dt + e_{1U} i_1 dt + e_{1L} i_1 dt$$
$$= i_1^2 R_1 dt + e_{1U}(i_0 + i_1')dt + e_{1L} i_1 dt$$

$$= i_1^2 R_1 dt + e_{1U} i_0 dt + e_{1U} i_1^{'} dt + e_{1L} i_1 dt$$
$$= i_1^2 R_1 dt + e_{1U} i_0 dt + e_2 i_2 dt + e_{1L} i_1 dt \qquad \text{(using Eqn.5.19)}$$

Of the right hand terms of this equation, the first one is heat in R_1. The integral of the second term $\frac{1}{2} L_{1U} {I_1}^2$ is the SE_M in L_{1U}. The third term is, by Eqn.A, the energy transferred to circuit 2. The integral of the fourth term $\int_{t=0}^{t=\infty} e_{1L} i_1 \, dt = \int_0^{\infty} L_{1L} \frac{di_1}{dt} i_1 \, dt = L_{1L} \int_0^{I_1} i_1 . di_1 = \frac{1}{2} L_{1L} {I_1}^2$ is the SE_M in L_{1L}. Hence, the complete energy equation during the entire transient period,

$$\int E_1 i_1 dt = \int i_1^2 R_1 dt + \frac{1}{2} L_{1U} {I_1}^2 + \int i_2^2 R_2 dt + \frac{1}{2} L_{1L} {I_1}^2$$
$$= \int i_1^2 R_1 dt + i_2^2 R_2 dt + \frac{1}{2} L_1 {I_1}^2$$

In words, *Battery output = Circuit 1 heat + Circuit 2 heat +* SE_M *in coil 1 total inductance.* In both the situations (a) and (b), during growth of I_1, coil 1 source E_1 supplies the SE_M in the magnetic field of coil 1 and the ohmic heat in both coils. After coil 1 current attains its final steady value I_1, all the transient quantities e_1, e_2 and i_2 decay to zero; E_1 continues to heat R_1, and no power or energy is there in coil 2. These observations are valid both for cumulative and differential coupling of the coils; in the latter case, the direction of e_2 and i_2 through coil 2 is reversed compared to the former case.

5.6.5 Energy stored in mutual inductance (coupled coils): Let us next insert a battery in circuit 2 of Fig.5.10. Neglecting magnetic leakage of the coils, Fig.5.11 shows the present circuit for cumulative coupling between the coils. Before the switch is closed, circuit 2 carries a steady current $I_2 = E_2/R_2$ (ignoring the cell e_2) in the direction shown, and this produces a common flux ϕ_{21} in the direction shown. As circuit 1 is then switched on, i_0 and φ grows and induces the emfs $e_1 = N_1 \frac{d\varphi}{dt} = L_1 \frac{di_0}{dt}$ in circuit 1,

and $e_2 = N_2 \frac{d\varphi}{dt} = M \frac{di_0}{dt}$ in circuit 2,

$\varphi = \varphi_{12} + \phi_{21} \rightarrow \phi_{12} + \phi_{21}$

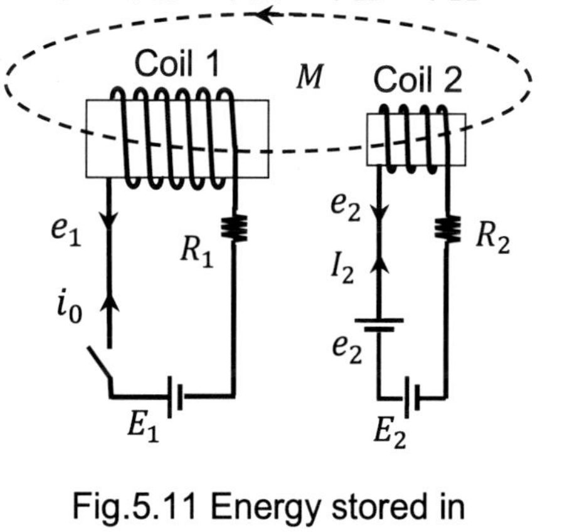

Fig.5.11 Energy stored in mutual inductance

in the directions shown, M being the mutual inductance between coils. To suppress the resulting circuit 2 current i_2, if we insert a cell e_2 in opposition to the induced emf e_2, as shown, circuit 2 current will be maintained at its steady value I_2. As such, circuit 1 current i_0 will grow exponentially to its steady value I_1, and the common flux $\varphi = \varphi_{12} + \phi_{21}$ to its steady value $\phi = \phi_{12} + \phi_{21}$, as we discussed in Sec.4.17.1.

During growth of circuit 1 current, battery E_1 supplies the $SE_M = \frac{1}{2}L_1I_1^{\ 2}$ and $\int i_0^2 R_1 dt$ energies, as if the coil 2 were not there. In circuit 2, as the current does not change, battery E_2 output power continues to be $I_2^{\ 2}R_2$, and the energy delivered by the cell e_2 during time dt of the growth process

$$e_2 I_2 dt = N_2 I_2 d\varphi = M I_2 di_0 = d(SE_M)_M \qquad \text{.............. (5.20(a))}$$

increases the energy stored in the mutual inductance M by the amount $d(SE_M)_M$. Total electromagnetic energy transferred from the cell e_2 to the mutual inductance, therefore,

$$(SE_M)_M = \int d(SE_M)_M = M I_2 \int_0^{I_1} di_0 = M I_1 I_2 \qquad \text{.............. (5.20)}$$

Circuit 2 also had a $SE_M = \frac{1}{2}L_2I_2^{\ 2}$ prior to energizing circuit 1. Total SE_M in the system of two cumulatively coupled coils is thus $SE_M = \frac{1}{2}L_1I_1^{\ 2} + \frac{1}{2}L_2I_2^{\ 2} + MI_1I_2$. If the coils be differentially coupled, the directions of the emf e_2 will be opposite compared to that of e_1. As such, the cell e_2, the polarity of which has also to be reversed, will now receive the energy MI_1I_2. So the SE_M remaining in the system would be $SE_M = \frac{1}{2}L_1I_1^{\ 2} + \frac{1}{2}L_2I_2^{\ 2} - MI_1I_2$. Combining these two equations, we write

$$SE_M = \frac{1}{2}L_1I_1^{\ 2} + \frac{1}{2}L_2I_2^{\ 2} \pm MI_1I_2 \qquad \text{......................... (5.21)}$$

(positive sign for cumulative coupling and negative for differential). SE_M is always positive, and to prove this we need to prove that $\frac{1}{2}L_1I_1^{\ 2} + \frac{1}{2}L_2I_2^{\ 2} > MI_1I_2$. By mathematics, the square of a difference is always positive.

So, $(\sqrt{L_1}I_1 - \sqrt{L_2}I_2)^2$ is positive

or, $L_1I_1^{\ 2} + L_2I_2^{\ 2} - 2\sqrt{L_1L_2}I_1I_2$ is positive

or, $\frac{1}{2}L_1I_1^{\ 2} + \frac{1}{2}L_2I_2^{\ 2} - \sqrt{L_1L_2}I_1I_2$ is positive

or, $\frac{1}{2}L_1I_1^{\ 2} + \frac{1}{2}L_2I_2^{\ 2} > \sqrt{L_1L_2}I_1I_2$

Eqn.5.13, ($M = k\sqrt{L_1L_2}$), reveals that M is smaller than $\sqrt{L_1L_2}$. Now the logic is, as the sum $\frac{1}{2}L_1I_1^{\ 2} + \frac{1}{2}L_2I_2^{\ 2}$ is greater than $\sqrt{L_1L_2}I_1I_2$, it is still greater than MI_1I_2, and hence the proof.

Eqn.5.21 can be expressed in the form

$$SE_M = \frac{1}{2}L_1I_1^{\ 2} + \frac{1}{2}L_2I_2^{\ 2} \pm MI_1I_2 = \frac{1}{2}(L_{1L}+L_{1U})I_1^{\ 2} + \frac{1}{2}(L_{2L}+L_{2U})I_2^{\ 2} \pm MI_1I_2$$

Now, $L_{1U} = \frac{N_1^{\ 2}}{S} = \frac{N_1N_2}{S}.\frac{N_1}{N_2} = M\frac{N_1}{N_2}$, and similarly, $L_{1U} = M\frac{N_2}{N_1}$. If $N_1 = N_2$, we will have

$$L_{1U} = L_{2U} = M$$

and the expression assumes the form

$$SE_M = \frac{1}{2}(L_{1L}+M)I_1^{\ 2} + \frac{1}{2}(L_{2L}+M)I_2^{\ 2} \pm MI_1I_2 \qquad \text{............... (5.21(a))}$$

This equation is analogous to the corresponding Eqn.1.44(a) in Electrostatics.

If the cell e_2 were not inserted, it can be shown that the energy stored in mutual inductance MI_1I_2 is shared by both the cells E_1 and E_2 and a portion is transferred by transformer principle. The cells deliver their shares for cumulative coupling of the coils, or they receive for differential coupling. The proof of this case remains an exercise for interested readers.

5.7 Mutual Induction by Dynamic Methods – Electromechanical Energy Conversion: As was stated in Sec.5.2, a dynamically induced emf is generated by mutual induction in an electrical conductor or circuit by its relative motion with respect to a neighbouring magnet or electromagnet. In Sec.4.17 (Force and Potential Energy of Current Carrying Coils), three common types of motion, (a) axial (b) traverse and (c) angular, were considered. As shown in Fig.5.1, same types of motion are also applicable in the present topic. Furthermore, three situations were considered in Sec.5.6 (Mutual Induction by Static Method) with Figs.5.6 (coil 2 circuit open), 5.10 (coil 2 circuit closed) and 5.11 (coil 2 circuit contains a source). Same set of situations will also be considered here under each type of motion. We will thus consider a total of nine situations in following three sub sections.

5.7.1 Dynamic induction with axial motion: In the figures in Fig.5.12, two cumulatively wound coils are in relative axial motion, suppose coil 1 is fixed and coil 2 moves. Also suppose any one coil, say coil 1, is an electromagnet with N_1I_1 ampere-turns, and coil 2 of N_2 turns moves in the field of coil 1. Same as in Sec.4.17.1, we further suppose that coil 2 does not possess any magnetic field, so experiences force and possesses PE_M by virtue of its presence in the field of coil 1. As coil 2 approaches coil 1 (moves towards left) with velocity v, or by distance dx in time dt, mutual flux from coil 1 linking coil 2 increases by $d\varphi_{12}$. This induces in coil 2 a flux linkage emf $e_2 = N_2.d\varphi_{12}/dt$, in the same way as in static induction, Fig.5.6. In fact, induced emf results from change in flux linkage, no matter how the change occurs, statically or dynamically. This flux linkage emf may also be viewed as flux cut or dynamic emf in the way the turns of coil 2 cuts the flux of coil 1; this explanation will become more convincing in next two sections. Shown directions of v and e_2 comply with Lenz's law; if the direction of v is reversed, that of e_2 also reverses.

Mutual induction by dynamic method, however, differs from the static method in one important aspect. We have in general, $M = N_2\phi_{12}/I_1$ (*mutual flux linkage of secondary per ampere in primary*), or $MI_1 = N_2\phi_{12}$. In dynamic method, as the distance x between the coils decreases by dx, both ϕ_{12} and hence M increase by $d\varphi_{12}$ and dM respectively, I_1 remaining constant. Hence,

$$N_2.d\varphi_{12} = dM.I_1 \quad \text{and}$$

$$N_2\frac{d\varphi_{12}}{dx} = \frac{dM}{dx}I_1$$

where, $d\varphi_{12}/dx$ and dM/dx both are negative. This should be compared with the static case, wherein M is constant and I_1 varies.

(a) In Fig.5.12(a), as coil 2 circuit is open, it does have any current, so does not experience any electromechanical force on its motion. This dynamic situation is analogous to the static case of Sec.5.6, and forms the principle of induced emf generation in electrical machines.

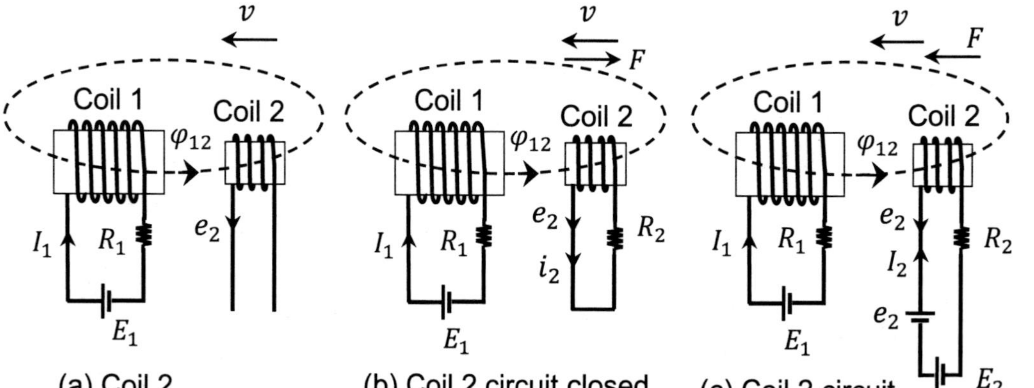

(a) Coil 2 circuit open (b) Coil 2 circuit closed (c) Coil 2 circuit contains a source

Fig.5.12 Dynamic induction with axial motion

(b) In Fig.5.12(b), coil 2 circuit is closed with its resistance R_2. Current i_2 generates a repulsive force F on coil 2. As $d\varphi_{12}$ is increase in opposing flux, work done by some mechanical agency in moving coil 2 against the generated force, as discussed in Sec.4.17.1, $F.dx = N_2 i_2 . d\varphi_{12}$. Also, energy generated in coil 2, $e_2 i_2 dt = N_2 i_2 . d\varphi_{12}$. Hence, mechanical energy delivered by the agency has been converted into electrical energy consumed in heating coil 2 circuit resistance. The result is independent of the direction of v. This situation is analogous to the static case of transformer principle, Sec.5.6.4, and forms the principle of electric generator.

(c) In Fig.5.12(c), coil 2 circuit contains a battery (source) E_2, so carries a current $I_2 = E_2/R_2$ in the direction shown, before any movement of the coil. This current I_2 generates an attractive force F on coil 2. If this attraction moves the coil, induced emf e_2 would cause some current i_2. To prevent i_2 to occur, we may insert a cell e_2 in opposition to the induced emf e_2, as shown, and as such, coil 2 current remains unaltered at I_2 during movement of the coil. As coil 2, with this constant current I_2, moves under attraction F, it delivers an mechanical output energy $F.dx = N_2 I_2 . d\varphi_{12}$, and its PE_M reduces by the same amount, $d\varphi_{12}$ being increase in helping flux now. Also, battery E_2 output power continues to be $I_2{}^2 R_2$ heat, and cell e_2 output energy $e_2 I_2 dt = N_2 I_2 . d\varphi_{12} = dM . I_1 I_2$ adds to the magnetic field as increase in its $(SE_M)_M$. Hence, we have the important results

Input electrical energy as cell e_2 output, $e_2 I_2 dt$
= Input to magnetic field as gain in its $(SE_M)_M$, $+d(SE_M)_M = +dMI_1 I_2$
= Output from magnetic field as loss in its PE_M, $-dPE_M = N_2 I_2 d\varphi_{12}$

$$= \text{Output mechanical energy, } Fdx \qquad \text{.................. (5.22)}$$

$$F = -\frac{dPE_M}{dx} = \frac{d(SE_M)_M}{dx} = \frac{dM}{dx} I_1 I_2 \qquad \text{.......................... (5.23)}$$

and also that positive change in PE_M is negative change (like mirror image) in $(SE_M)_M$, energy stored in mutual inductance between two current carrying coils, so cannot occur with any single coil, in line with our previous discussions in Sections 1.8 and 3.8.

This dynamic situation is analogous to the static case of Sec.5.6.5.

If cell e_2 is not inserted, coil 2 current will be reduced from its standstill value I_2 to $i = (I_2 - i_2)$ in the direction of I_2. Voltage equation of circuit 2 will be $E_2 = iR_2 + e_2$, and energy equation

$$\begin{aligned} E_2 idt &= i^2 R_2 dt + e_2 idt \\ &= i^2 R_2 dt + N_2 i.d\varphi_{12} \\ &= i^2 R_2 dt + F.dx \end{aligned}$$

That is, electrical energy supplied by battery $E_2 =$ heat loss in $R_2 +$ mechanical output. This forms the principle of electric motor.

As the reverse process, if two cumulatively coupled current carrying coils are pulled apart against attraction, or two differentially coupled current carrying coils are pushed closer against repulsion, by some external agency, mechanical energy is converted into electrical energy. Two conditions may again arise. If a reverse cell e_2 is inserted against induced emf e_2, the converted electrical energy will be fed back to this cell. If the reverse cell is not inserted, coil 2 current will become $i = (I_2 + i_2)$ and the converted electrical energy will add to the enhanced battery E_2 output electrical energy in overheating R_2, causing possible burning out of coil 2. This latter situation resembles to rotating a motor (of which the armature or rotor is coil 2) against its developed torque by some mechanical agency.

Such conversion of electrical energy into mechanical energy, and the reverse process, through electromagnetic induction are collectively referred to as *electromechanical energy conversion*.

It should be noticed in the above discussion that coil 1 (electromagnet) has no role other than to provide a constant flux ϕ. Hence, the above concepts also remain valid if the electromagnet be replaced by a permanent magnet or an isolated magnetic pole. Only difference is that, in the latter case mutual inductance M, and hence $(SE_M)_M$, are not defined.

5.7.2 Dynamic induction with traverse motion: Concepts of dynamic induction have been discussed in detail with reference to axial motion in previous section 5.7.1. It was seen in Ch.4 that whereas rotary motion presents the practical situation in rotating machines, traverse motion represents its linear or developed model. Hence, the derivations in present two sections follow from the previous section. Further, the PE_M to $(SE_M)_M$ relationships described by Eqns. 5.22 and 5.23, remain implied in present sections.

(a) In Fig.5.13(a), as the conductor AB is moved across the magnetic field, it cuts the magnetic flux of the field, and dynamically induced emf is generated in the conductor. Let the magnetic field be uniform, of flux density B, produced by N_1I_1 ampere turns (not shown in the figure), and l be the active length of the conductor in the field. If the conductor is moved by distance dx in time dt with velocity $v = dx/dt$, the amount of field flux the conductor cuts $d\varphi_{12} = Bldx$, and the flux cut emf, $e_2 = d\varphi_{12}/dt = Bl.dx/dt = Blv$. For one coil side composed of N_2 turns of conductors like AB, the flux cut emf

$$e_2 = N_2Blv \qquad \text{.............................. (5.24)}$$

'*Fleming's Right Hand Rule*' depicted in Fig.5.13(b) gives the direction of flux cut emf. In the figure, the conductor is moved towards left, and the flux cut emf is downwards the conductor. As electric generators operate on this principle '*an induced emf is generated in a conductor as it cuts (or is cut by) magnetic flux*', we commonly refer it as '*generator principle*'. Fleming's right hand rule that gives the direction of flux cut emf, with respect to the directions of the magnetic field and the motion, is called the '*generator rule*'. Fleming's right hand rule, and the left hand rule explained earlier in Sec.4.12, are reproduced side by side at Fig.5.13(c) for comparison. Following similarities or tips may help remember these rules together.

Forefinger : Magnetic quantity – Magnetic field or flux
Middle finger: Electrical quantity – Current in LHR (*cause*) and emf in RHR (*effect*)
Thumb : Mechanical quantity – Force in LHR (*effect*) and velocity in RHR (*cause*)

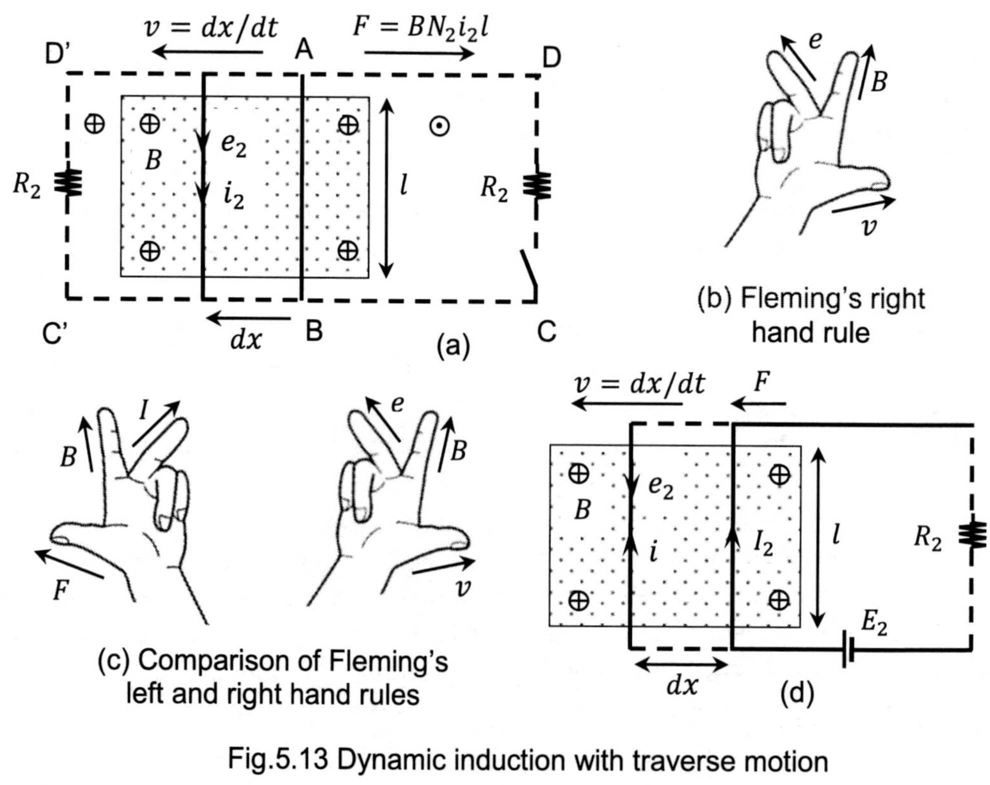

Fig.5.13 Dynamic induction with traverse motion

As was mentioned in Sec.5.2, flux cut principle can always be explained in terms of flux linkage principle. The active length of the conductor AB considered in Fig.5.13(a) must be a portion of some electric circuit like ABCD. As this circuit or coil is moved by a distance dx towards left, the amount of flux linking the coil increases by $d\varphi_{12} = Bldx$, which is also the increase in the amount of flux cut. Hence the subject is explained.

In Fig.5.13(a), if the conductor circuit were closed, e_2 would cause i_2 in the shown direction as given by Fleming's right hand rule. The force on the conductor due to this current i_2 would, by Fleming's left hand rule, be towards right, which is opposite to the motion of the conductor towards left. The flux cut emf e_2 is thus so directed as to oppose the motion of the conductor to which it is due. Fleming's right hand rule, therefore, complies with the fundamental principle of Lenz's law. This compliance can also be ascertained with flux linkage principle. As the conductor AB is moved towards left, the flux φ_{12} linking the coil ABCD increases. The magnetic field of the anticlockwise induced current i_2 round the coil acts in opposition to that of the main field, so opposes φ_{12} to increase. Also, if the coil ABC'D' be considered, the flux φ_{12} linking the coil decreases as the conductor is moved towards left, and the clockwise induced current i_2 round the coil acts in conjunction with that of the main field, so opposes φ_{12} to decrease. Hence, Lenz's law is complied.

(b) If the coil circuit ABCD be closed with its resistance R_2, emf $e_2 = N_2Blv$ causes $i_2 = e_2/R_2$ in the circuit. Force on the N_2- turn coil side due to this current, $F = BN_2i_2l$ acts towards right. Work done by some mechanical agency in moving the coil against F through distance dx towards left in time dt

$$F.dx = BN_2i_2ldx = N_2Blv.dt.i_2 = e_2i_2dt$$

is converted into the electrical energy consumed by R_2 in the coil circuit during time dt, and the result is independent of the direction of v.

(c) In Fig.5.13(d), the coil circuit contains a battery E_2, so carries a current $I_2 = E_2/R_2$ in the direction shown, and experiences a force F directed along left. If this force F moves the coil by distance dx in time dt with velocity $v = dx/dt$, dynamic emf $e_2 = N_2Blv$ and induced current $i_2 = e_2/R_2$ will be generated in the coil circuit, as shown in Fig.(a). As i_2 directs against I_2, net current along I_2 becomes $i = (I_2 - i_2)$, and this gives $F = BN_2il$.

Mechanical output from the system, which is the work done by the field, in time dt,

$$F.dx = BN_2ildx = N_2Blv.dt.i = e_2idt$$

Voltage equation of coil circuit, $E_2 = iR_2 + e_2$, and the energy equation for time dt

$$E_2idt = i^2R_2dt + e_2id = i^2R_2dt + F.dx$$

In words, electrical energy supplied by battery E_2 = heat loss in R_2

+ mechanical energy obtained by electromechanical energy conversion.

Mutual inductance between the field and the coil ABCD, $M = N_2\phi_{12}/I_1$. As the coil moves by distance dx towards left (dx negative), mutual flux ϕ_{12} increases by $d\varphi_{12}$, and M increases by $dM = N_2d\varphi_{12}/I_1$.

Hence we have

$$N_2 \frac{d\varphi_{12}}{dx} = \frac{dM}{dx} I_1, \text{ where } d\varphi_{12}/dx \text{ and } dM/dx \text{ both are negative.}$$

As such, we write the force (in the direction of motion) as

$$F = BN_2 il = N_2 \frac{Bl.dx}{dx} i = N_2 i \frac{d\varphi_{12}}{dx} = \frac{dM}{dx} I_1 i$$

which is Eqn.5.23.

5.7.3 Dynamic induction with rotary motion:

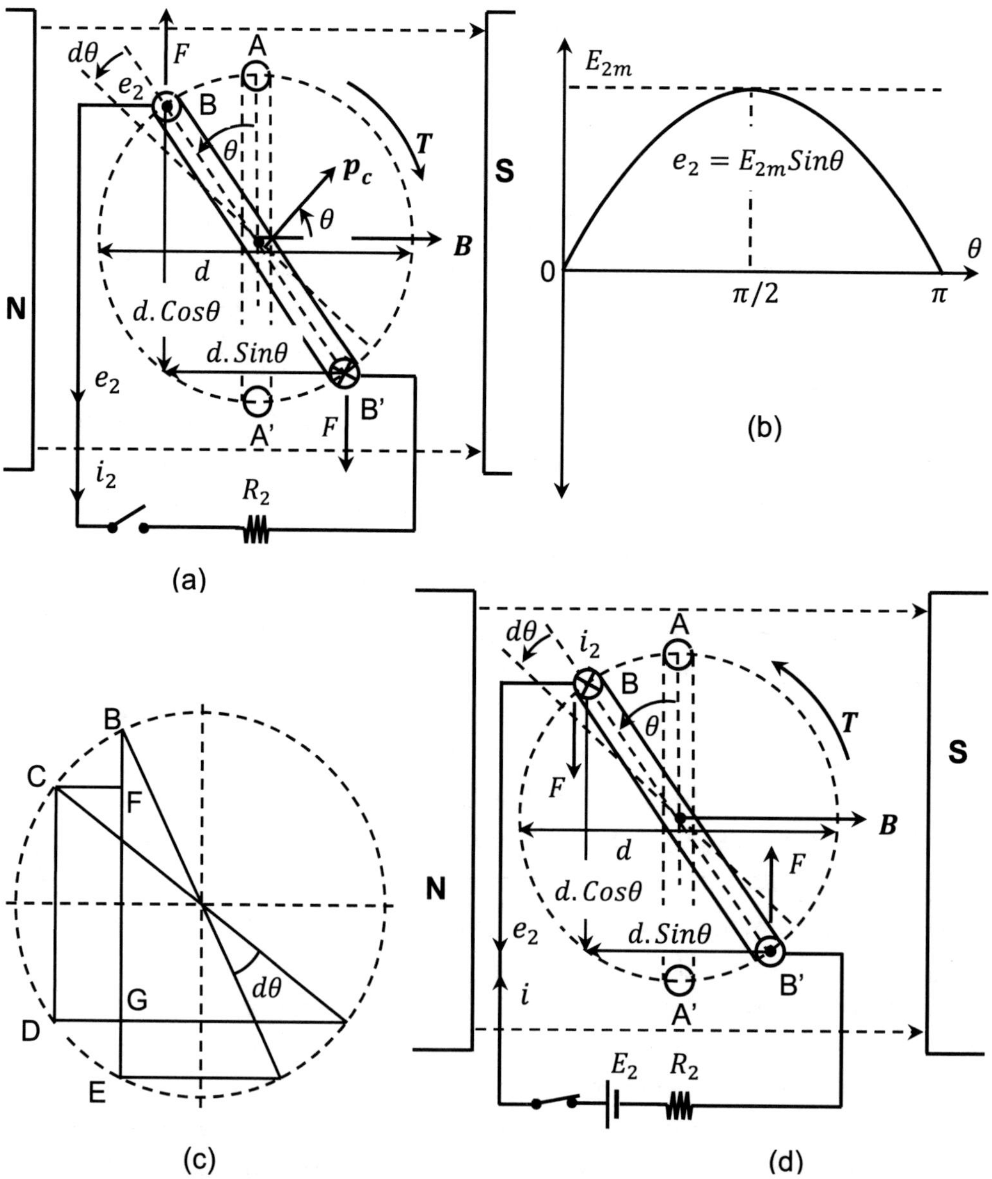

Fig.5.14 Dynamic induction with rotary motion

The system shown in Fig.5.14 is reproduction of those in Figs.4.14 and 4.22. BB' is one advancing position of the rectangular coil 2 rotated along anticlockwise or positive direction from the zero reference or starting position AA'. The coil has N_2 number of turns, width d and active length l in the uniform magnetic field B produced by $N_1 I_1$ ampere turns (not shown in the figure).

(a) In Fig.5.14(a), the coil is open circuited. As it is rotated in anticlockwise direction through angle $d\theta$ in time dt, its coil sides move with a tangential linear velocity v. Flux cut emf induced in each coil side is $N_2 Blv Sin\theta$ in the shown directions given by Fleming's right hand rule.

Flux cut emf in the coil, as shown graphically in Fig.5.14(b),

$$e_2 = 2N_2 Blv Sin\theta = E_{2m} Sin\theta \qquad \text{.........................} (5.25)$$

This is the general form of emf equation in electrical machines.

To explain the flux cut emf in terms of flux linkage emf in this case, we consider the geometry of the system shown enlarged at Fig.4.14(c). As the angle $d\theta$ is infinitesimal,

Arc BC = Chord BC

or, $r.d\theta = v.dt$ ($r = d/2$ being the radius of rotation of the coil)

and, BF = Chord BC.$Sin\theta = v.dt.Sin\theta$.

As the coil rotates through $d\theta$ in time dt, its flux linkage decreases by

$$d\varphi_{12} = Bl(\text{BE}-\text{FG}) = 2Bl.\text{BF}$$

$$= 2Blv Sin\theta dt = 2Blr Sin\theta d\theta = Bld Sin\theta d\theta \qquad \text{........} (A)$$

Hence, the flux linkage emf $e_2 = N_2 d\varphi_{12}/dt = 2N_2 Blv Sin\theta$, which is also the magnitude of the flux cut emf given by Eqn.5.25. Further, if the coil circuit were closed, e_2 would cause i_2 in the shown direction as given by Fleming's right hand rule. This would generate a clockwise torque, as shown, opposing the anticlockwise rotation of the coil to which the induced emf is due. Hence, Fleming's right hand rule is in compliance with Lenz's law.

(b) If the coil circuit be closed with its resistance R_2, emf $e_2 = 2N_2 Blv Sin\theta$ causes $i_2 = e_2/R_2$ in the coil. Clockwise torque on the coil due to this current, $T = BN_2 i_2 ld Sin\theta$. Work done by some mechanical agency in rotating the coil against T through the angle $d\theta$ in time dt

$$T d\theta = BN_2 i_2 ld Sin\theta d\theta = (2N_2 Blv Sin\theta) i_2.dt = e_2 i_2 dt$$

is converted into the electrical energy consumed by R_2 in the coil circuit during time dt, and the result is independent of the direction of rotation of the coil.

This is the process of electromechanical energy conversion in an electric generator, on which R_2 is the load.

(c) In Fig.5.14(d), the coil circuit contains a battery E_2, so carries a current $I_2 = E_2/R_2$ in the direction shown, and the coil experiences an anticlockwise torque T. If this torque rotates the coil through angle $d\theta$ in time dt with linear velocity v, a dynamic emf $e_2 = 2N_2 Blv Sin\theta$ and a current $i_2 = e_2/R_2$ will be generated in the coil circuit, as shown in Fig.(a). As i_2 directs against I_2, the net current along I_2 becomes $i = (I_2 - i_2)$ and the torque becomes $T = BN_2 ildSin\theta$. Mechanical output from the system, which is the work done by the field, in time dt,

$$Td\theta = BN_2 ildSin\theta d\theta = (2N_2 BlvSin\theta)i.dt = e_2 idt$$

Voltage equation of coil circuit is $E_2 = iR_2 + e_2$, and the energy equation for time dt

$$E_2 idt = i^2 R_2 dt + e_2 idt \ = i^2 R_2 dt + Td\theta$$

In words, electrical energy supplied by battery E_2 = heat loss in R_2

+ mechanical energy obtained by electromechanical energy conversion.

This is how electromechanical energy conversion occurs in electric motors.

Mutual inductance between the field and the coil, $M = N_2 \phi_{12}/I_1$. As the coil rotates in positive or anticlockwise direction through angle $d\theta$ ($d\theta$ positive), mutual flux ϕ_{12} decreases by $d\varphi_{12}$, and M decreases by $dM = N_2 d\varphi_{12}/I_1$. Hence we have

$N_2 \frac{d\varphi_{12}}{d\theta} = \frac{dM}{d\theta} I_1$, where $d\varphi_{12}/d\theta$ and $dM/d\theta$ both are negative.

As Eqn.(A) provides $BldSin\theta = \frac{d\varphi_{12}}{d\theta}$, we write the torque (in the direction of rotation) as

$$T = BN_2 ildSin\theta = N_2 \frac{d\varphi_{12}}{d\theta} i$$

$$= \frac{dM}{d\theta} I_1 i \qquad \text{..} \ (5.26)$$

This relationship forms the analytical part of the *Generalised Theory of Electrical Machines*.

Ch. 6: Magnetic Hysteresis and Permanent Magnets

6.1 Magnetic Hysteresis:

6.1.1 Hysteresis loop and its terminology:

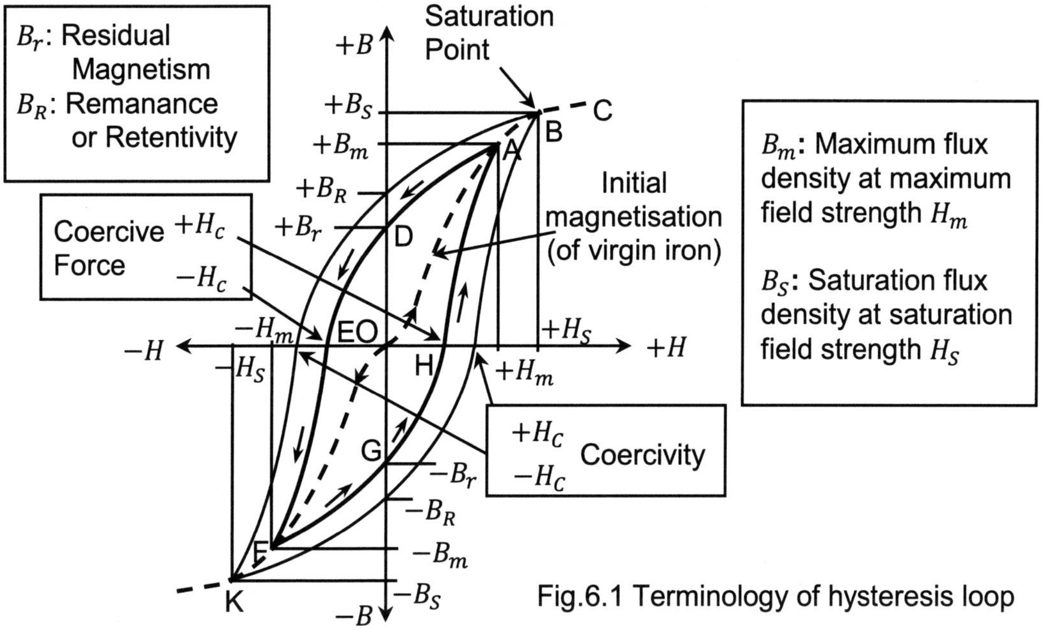

Fig.6.1 Terminology of hysteresis loop

In Sec.3.12.2 (Fig.3.16) it was shown how the flux density B in a virgin magnetic material increases with increasing values of the external magnetic field H. The B-H curve of Fig.3.16 is reproduced here as the dotted line OABC in Fig.6.1. After B attains some maximum value B_m at some maximum value H_m of the field strength, if H is then reduced, B although reduces but does not follow its ascending curve. Instead, it descends following the curve AD, and this descending curve can be seen to lie above the ascending curve. When H has been reduced to zero, point D on the curve, the flux density OD remaining in the iron is called the *Residual* or *Remanent Flux Density* or *Residual Magnetism*, B_r. If H is then increased in the reverse direction, B reduces along the curve DE. The value of the reverse field strength OE required to wipe out the residual flux density, point E on the curve, is called the *Coercive Force*, H_c. The values of B_r and H_c depend on the value of H_m, on the quality of the iron, and on the possible amount of air gap in the magnetic circuit. Present discussion with reference to Fig.6.1 remains for a closed magnetic circuit of iron, *without any air gap.* Effect of air gap will be discussed in Sec. 6.3. Further increase of H in the reverse direction causes B to increase in the reverse direction along the curve EF. If the magnitude of the reverse field strength is increased upto the same maximum value as it was in the forward direction, point F on the curve, the same maximum flux density is obtained in the reverse direction as it was in the forward direction. If H is next increased in the forward direction from negative maximum $-H_m$, via zero, back to the positive maximum $+H_m$, the flux density traces the curve FGHA which is symmetrical to the curve ADEF. It can be seen that the zero values of B occur *later* than those of H in both directions of magnetisation. In other words, the flux density 'lags' behind the field strength producing it. This peculiar property of lagging of the flux density in a magnetic material behind the magnetic field strength producing it is referred to as *Magnetic Hysteresis*. The closed loop of curve ADEFGHA traced out by the B and H values during the complete cycle of magnetisation and demagnetisation is called the *Hysteresis Loop*. For any given specimen of iron, the hysteresis loops with higher values of H_m will certainly become larger in size. If an iron with closed magnetic circuit is magnetised upto saturation level, the amount of residual flux density obtained is termed as *Remanence* or *Retentivity* B_R, and the amount of coercive force required to wipe out the remanence flux density as the *Coercivity* H_C of the material. The remanence and coercivity are, therefore, properties of magnetic materials. For any given material, the largest possible hysteresis loop is, of course, that obtained with saturation level of the flux density at both directions ($\pm B_S$ in Fig.6.1). This is called the *major loop*, which is unique for the material. Inside a major loop there may be innumerable number of minor loops. Such minor loops, as will be explained in the following section, may be symmetrical or asymmetrical. The dotted virgin curve shown in Fig.6.1 for the initial magnetisation is also referred to as the *normal magnetisation curve*; 'normal' simply denotes the absence of hysteresis. Principal demerit of hysteresis lies in that it causes energy loss in the form of heating the iron used in electrical equipment. It has several merits as well. For example, permanent magnets would not exist, or DC self-excited shunt generators would not build up voltage, had there been no hysteresis.

6.1.2 Features of hysteresis loops:

(a) Cyclic state: It appears from the above discussion that as H is repeatedly varied between $\pm H_m$, the same hysteresis loop ADEFGHA will be reproduced. Such a situation is referred to as '*cyclic state*'. A piece of iron is said to be in a cyclic state if it reproduces the same hysteresis loop when taken through repeated cycles of magnetisation and demagnetisation with the same value of $\pm H_m$. Such a situation, however, does not occur ordinarily. When a virgin piece of iron is magnetised for the first time (or some pre-magnetised piece is re-magnetised with some different value of $\pm H_m$), the successive hysteresis loops for the first few cycles will considerably differ from one another. The cyclic state in a piece of iron can be easily brought by subjecting it to repeated cycles of magnetisation and demagnetisation for some fifteen to twenty times with gradually increasing values of the external field upto the desired value $\pm H_m$.

(b) Restoration of the virgin state: Once a virgin piece of iron is magnetised, its virginity is lost, in the sense the flux density does not become zero with zero field strength. The virginity so lost due to hysteresis can, however, be restored. Restoration of virginity simply means wiping out of residual magnetism, and is usually referred to as '*demagnetisation*'. There are several means for demagnetisation. One simple way to achieve demagnetisation is to subject the specimen to repeated cycles of magnetisation and demagnetisation with gradually reducing values of the external field to zero. This process of demagnetisation is thus the reverse process of achieving cyclic state.

(c) Mid course reversals: In Fig.6.2, the hysteresis loop 12 is symmetrical about the B-H axes. If H is increased from some point like 3 midway the descending curve 1432, B increases along the curve 356 merging with the descending curve at point 4. If H is reversed from point 5 or 6, B will back to point 3. If H is increased from point 7, B will move to point 6. Similarly, if points like 8 and 9 are considered, curves like 801 and 91 will be traced out, and so on. Similar set of curves will also be there for the ascending curve 21. In spite of the variety of hysteresis characteristics exhibited by different magnetic materials, some general regularity is observed. These regularities were identified in the early twentieth century as a set of five rules known as Madelung's rules. Although it is not necessary now to enter into the details of these rules, it can be said that the present discussion is in conformance to these rules. A number of inner loops have been shown in Fig.6.2 inside the outer loop 12, but all the outer and inner loops are minor loops only. If H_m on positive and negative sides are not equal, a loop like 19 will be formed. Such a situation is usually uncommon in power engineering, but may occur if the magnetising winding carries an alternating current superimposed on a direct current, as, for example, one winding of a radio transformer. The loop 19 is not symmetrical about the B-H rectangular axes, but unless saturation is approached to, it is almost symmetrical about some displaced set of X-Y perpendicular axes, as shown. If, however, H_m enters the saturation zone on any one side, the resulting hysteresis loop not only becomes unsymmetrical about any set of perpendicular axes, but also gets considerably distorted in shape. Such unsymmetrical loops are seldom encountered in electrical power equipment.

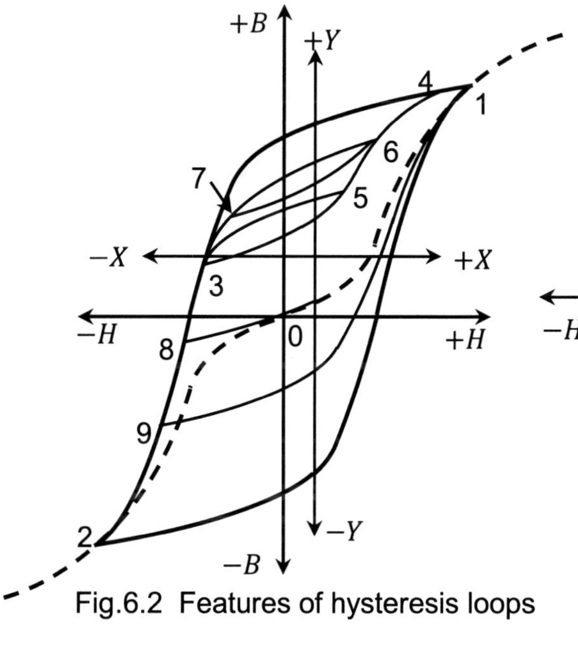

Fig.6.2 Features of hysteresis loops

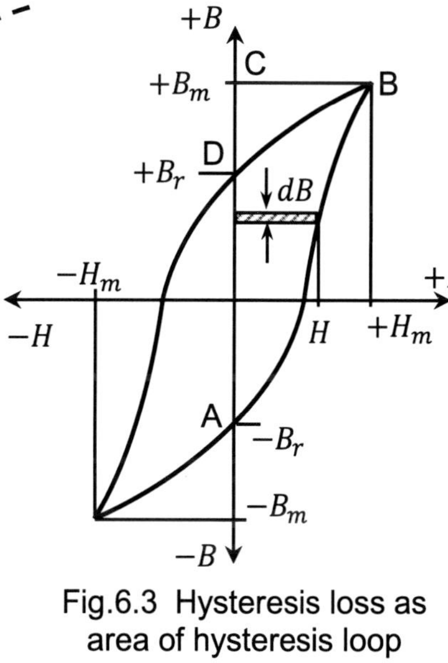

Fig.6.3 Hysteresis loss as area of hysteresis loop

6.1.3 Recoil loops, lines and recoil permeability: The small inner loops like 35 and 76 in Fig.6.2 are called *'recoil loops'*. Such recoil loops are so narrow that it is sufficient to represent them by straight lines like 35 and 76 respectively, called *'recoil lines'*. For symmetrical outer loop, the demagnetisation curve 2841 can be seen to have an initial portion, a linear portion, a knee portion and a saturation portion, similar to those on the normal magnetisation curve 01.

The slope of the recoil lines originating from the linear portion of the demagnetisation curve is constant for any material, and this defines one permeability (property) of the material termed its *recoil permeability* as

$$\mu_{recoil} = \mu_{rrecoil} \cdot \mu_0$$
$$= \text{Average } \Delta B/\Delta H \text{ slope of the recoil lines} \quad \text{........ (6.1)}$$

where, $\mu_{rrecoil}$ is the *relative recoil permeability*. Recoil permeability of a material is decided by the shape of its symmetrical major loop, and in any case it is certain to be lower than its conventional permeability given by the straight line portion of the normal magnetisation curve. As the originating point climbs up the knee portion of the demagnetisation curve, the slope of the recoil lines reduces. Beyond saturation point, the major loop ceases to a straight line of air permeability slope μ_0.

With unsymmetrical outer loops associated with an AC component superposed on the DC magnetic field, recoil permeability deviates from constancy. In such situations, *incremental permeability* is defined as

$$\mu\Delta = \Delta B/\Delta H \quad \text{.. (6.2)}$$

where ΔB and ΔH refer to a *small* recoil loop at any point on the outer loop.

As was mentioned above, incremental permeability is not of much significance in electrical power engineering. Moreover, topics like *intrinsic hysteresis loop*, related terminology and many other permeability definitions, which physicists are mainly interested in, have been excluded from our present scope of discussion.

6.1.4 Hysteresis loss: Magnetic materials used in most electrical equipment undergo cycles of magnetisation and demagnetisation continuously at some frequency of the system. Owing to hysteresis effect, such cycles of magnetisation and demagnetisation involve expenditure of energy due to 'molecular magnetic friction' in the material. The energy so expended is referred to as *hysteresis loss*, and is dissipated in heating the material. It was discussed in Sec.3.14 that whenever a magnetic field is created, it stores some magnetic potential energy. This energy must be drawn from the source that creates the field, and the source is obviously the electric circuit that produces current for the field. If the current is then reduced, the stored energy is returned back to the source. This principle is utilised in deriving the hysteresis loss.

Let us consider a closed magnetic circuit of axial length l and uniform cross sectional area A. As was discussed in Sec.4.6, a current i through a N-turn magnetising winding produces an axial field strength $H = Ni/l$, flux ϕ and flux density $B = \phi/A$ in the iron core. If the current is increased by di during time dt, the flux will increase by $d\varphi$, and an electromagnetic emf, Sec.5.4, $e = N.d\varphi/dt$ will be induced in the electric circuit. Energy transferred from the electric circuit to the magnetic circuit in time dt becomes

$$ei.dt = Ni.d\varphi = Hl.A.dB = V.H.dB$$

where $V = lA$ is the volume of the magnetic circuit. Thus the energy stored per unit volume of the magnetic circuit increases by $H.dB$, as represented by the shaded elemental area in the hysteresis loop of the iron shown at Fig.6.3. As the field strength increases from zero to $+H_m$, the area ABC represents the increase in stored energy per unit volume. As the field strength is then reduced from $+H_m$ to zero, decrease in stored energy is represented by the area BCD. Hence, the energy lost in hysteresis is given by the difference area ABDA which is the right half of the hysteresis loop about the B-axis. Similarly, as H reduces from zero to $-H_m$ and then increases back from $-H_m$ to zero, the left half area of the hysteresis loop about the B-axis gives the hysteresis loss. Thus the hysteresis loss per unit volume of the material for one complete cycle of magnetisation and demagnetisation is given by the area of the hysteresis loop. The area of the loop, of course, has to be measured to scale. If 1 cm represents h ampere/metre on H-axis and b tesla on B-axis, and the loop measures A cm^2, the hysteresis loss in joule per cycle per cubic metre of the material becomes

$$W_h = A.h.b \qquad \text{...} (6.3)$$

This expression for hysteresis loss in terms of the area of hysteresis loop obviously applies both for symmetrical and unsymmetrical loops.

Another empirical formula for W_h for symmetrical hysteresis loops is due to Charles Proteus Steinmetz (1865-1923), an electrical engineer of U.S.A. This is referred to as *Steinmetz's Hysteresis Law*, expressed as

$$W_h = kB_m{}^x \qquad \text{...} (6.4)$$

This law applies under certain limits. The factor k is a constant for any magnetic material, called its *hysteresis coefficient*. For common magnetic materials, its value lies in the range 200-500. The value of the index x depends on the value of B_m. It generally lies in the range 1.5-2.5, but values as high as 3.5 have also been obtained. For usual values of B_m lying in the range 0.1-1.2 T in most applications, x can be taken as 1.6 with sufficient accuracy for most practical purposes. As such, Eqn.6.4 approximates to

$$W_h = kB_m{}^{1.6} \qquad \text{...} (6.5)$$

For values of B_m both below 0.1 T and above 1.2 T, x becomes greater than 1.6.

6.2 Soft and Hard Magnetic Materials: Relative shapes of hysteresis loops classify magnetic materials into two groups for distinct applications. These are a) Non-retentive or *soft* materials, and b) Retentive or *hard* materials. Fig.6.4 shows the comparative hysteresis loops of these two types of materials. Soft materials are characterised by long and narrow loops with low coercivity (below 100 A/m). These materials possess high permeability, as was mentioned in Sec.3.2, so can be easily magnetised, and demagnetised also, with minimum external field H. Further, narrow loop means low hysteresis loss. These favourable properties render soft materials suitable for use as electromagnets in variety of electrical machineries and apparatus. Hard materials, on the other hand, have short and wide hysteresis loops with high coercivity. The permeability is low and hence these materials can be magnetised only with difficulty, but once magnetised their magnetism cannot be destroyed easily. These essential requirements make hard materials suitable for permanent magnets, wherein the question of hysteresis loss does not arise.

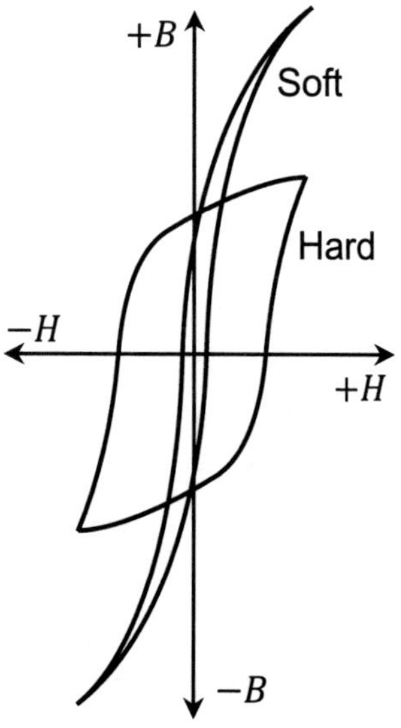

Fig.6.4 Soft and hard magnetic materials

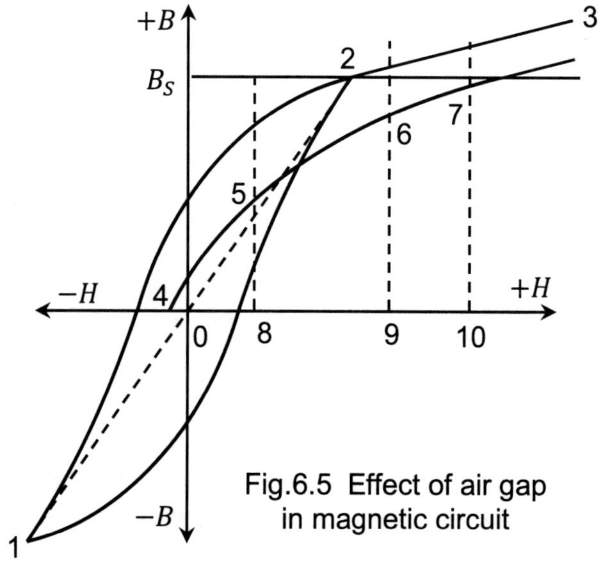

Fig.6.5 Effect of air gap in magnetic circuit

6.3 Effective Permeability in AC and DC Applications: Permeability of an iron is simply a measure how its flux density changes with change in the field strength. As hard materials are magnetised once during development of permanent magnets, and not during subsequent use, permeability is not an important quantity for hard materials. On the other hand, permeability is one prime quantity for soft materials which are repeatedly magnetised and demagnetised in regular course of use. This confines our discussion on permeability to soft materials only.

In Fig.6.5, 1-2 is a hysteresis loop, say major one with saturation flux density B_S, of a soft iron specimen *without any air gap*, as prevails in many alternating current applications like the core of a transformer or any choking coil. During operation, the permeability varies corresponding to all points on the hysteresis loop between the positive and negative peaks. The average (or effective) permeability is, therefore, simply the slope of the line 12 passing through the centre 0 of the B-H axes.

An air gap in magnetic circuit is an essential requirement in many electrical equipment. In Fig.6.5, 4-5-6-7 is portion of the demagnetising curve of the same specimen of iron, but *with a narrow air gap*. If this specimen is subjected to alternating field, as prevails in an induction motor for example, of maximum strength say 0-9, the straight line 0-6 gives the effective permeability. An air gap thus reduces effective AC permeability, which is quite logical.

Let us now consider direct current applications. Magnetic circuits in the field of DC machines contain air gaps. Thus, if the field current varies corresponding to field strengths between 0-8 and 0-9, the effective permeability is the slope of the recoil line 56. Similarly, if the current varies corresponding to field strengths between 0-9 and 0-10, the effective permeability is the slope of the recoil line 67. Were there no air gap in this latter case, the effective permeability would be air permeability μ_0 (slope of the line 23) which is less than the slope of the recoil line 67. Interestingly, a minor air gap increased the effective DC permeability in this case, as is commonly employed in the choking coil for smoothing the output current of rectifier, wherein a high inductance is desired, not a high flux density.

As the inductance of a coil or winding is proportional to the permeability of its core material,

$$L = \frac{N^2}{S} = \frac{N^2 \mu A}{l} \propto \mu \quad \text{(Sec.5.4)}$$

both these quantities undergo identical variations, other factors remaining same.

6.4 Theory of Permanent Magnets:

6.4.1 Conditions in permanent magnets: Fig.6.6(a) shows a specimen of some hard magnet steel, magnetised with its soft iron pole shoe, to form a permanent magnet. The useful air gap of the magnet is obtained on subsequent removal of the pole shoe. The magnetising winding on the magnet is not shown in the figure.

Let,
l_m = length of the magnet
a_m = cross-sectional area of the magnet
μ_m = permeability of the hard steel (magnet)
l_g = length of the pole shoe or of the air gap
a_g = cross-sectional area of the pole shoe or of the air gap
μ_i = permeability of the soft iron (pole shoe)

The reluctance of the magnet, pole shoe and the air gap are respectively given by

$$S_m = \frac{l_m}{\mu_m . a_m}$$

$$S_i = \frac{l_g}{\mu_i . a_g}$$

and $$S_g = \frac{l_g}{\mu_0 . a_g}$$

As l_g is small compared to l_m and μ_i is too large compared to μ_m, S_i is insignificant compared to S_m. Against this, as μ_m is large compared to μ_0, S_g is quite significant compared to S_m. Curve A in Fig.6.6(b) shows the demagnetisation portion of the hysteresis loop in the first and second quadrants for the magnet steel. Under the assumption $S_m \gg S_i$, the same curve A also applies to the specimen of the steel, with the pole shoe in position, for the same level of magnetisation. Further, let the level of magnetisation be the saturation level so that OD represents the remanence B_R and OE the coercivity H_C of the steel.

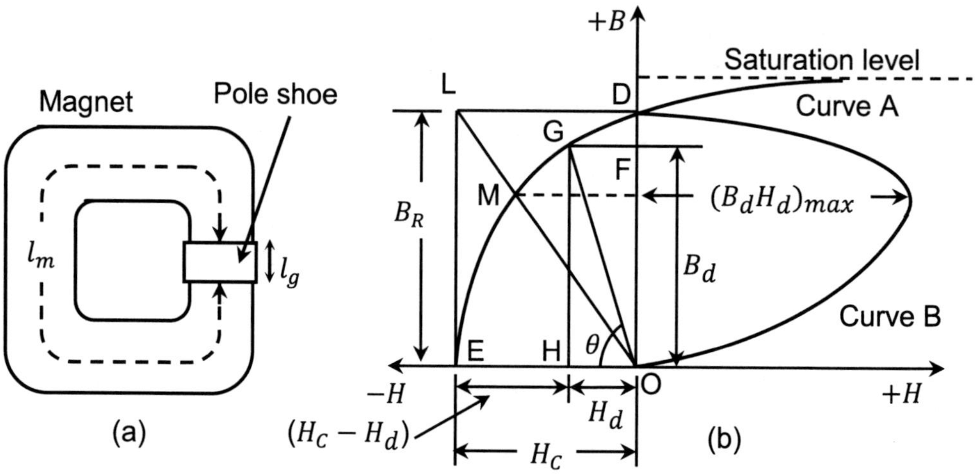

Fig. 6.6 Theory of permanent magnets

With the pole shoe in position, the flux density B_R in the magnet in the absence of the external field ($H = 0$) may be thought of as caused by the magnetising field H_C. Hence, the magnetic circuit equation (*flux = mmf / reluctance*) can be written as

$$B_R a_m = \frac{H_C l_m}{S_m}$$

If the pole shoe is then removed, the flux density in the magnet drops, due to the air gap, from OD($= B_R$), to some value OF($= B_d$). The magnetic circuit is now a series circuit of the magnet and the air gap, wherein the total mmf is $H_C . l_m$ inherent in the magnet. As such, the magnetic circuit equation (*mmf = flux x reluctance*) can be written in the form

$$H_C l_m = (B_d a_m) S_m + (B_g a_g) S_g$$

$$\text{or, } B_d a_m = \frac{H_C l_m - (B_g a_g) S_g}{S_m} \qquad \text{.. (A)}$$

where B_g is the flux density in the air gap. The flux density B_d in the magnet with the air gap could also be obtained with the pole shoe in position, point G on curve A, if a demagnetising field H_d were applied. As such, out of the total mmf $H_C . l_m$, the portion $(H_C - H_d) l_m$ is used in the magnet and the remaining portion $H_d . l_m$ in the air gap. In this case, the magnetic circuit equation for the flux in the magnet can be written as

$$B_d a_m = \frac{H_C l_m - H_d l_m}{S_m} \qquad \text{.. (B)}$$

Comparison of the above Eqns.(A) and (B) yields

$$H_d l_m = (B_g a_g) S_g$$

$$= B_g a_g \frac{l_g}{\mu_0 a_g}$$

$$= \frac{B_g l_g}{\mu_0} \qquad \text{.. (C)}$$

The situation is analogous to the fact that the current in an electric circuit would remain unaltered if a portion of the circuit resistance be replaced by an equivalent back emf. Thus, in Fig.6.7(a),

$$I = (E - I.r)/R$$

and in Fig.6.7(b), $I = (E - e)/R$

so that $e = I.r$

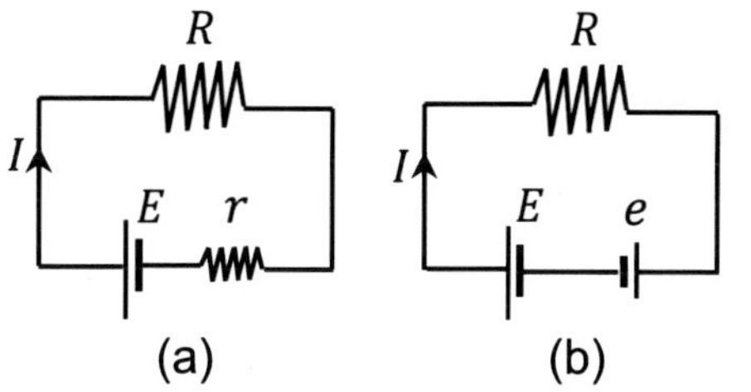

Fig.6.7 Analogous electric circuits

Although the actual operating point of the magnet with air gap is point F on the B-axis, the theory of permanent magnets is conveniently developed by assuming the operating point as G on curve A.

Neglecting magnetic leakage (Sec.4.7.4), the flux in magnet equals the flux in the air gap.

$$B_d a_m = B_g a_g \qquad \text{.......................................} (D)$$

Eqns. (C) and (D) form the two base equations from which the rest of the permanent magnet theory is derived.

6.4.2 Load Line of permanent magnet: Dividing Eqn.(D) by Eqn.(C) we get,

$$\frac{B_d\, a_m}{H_d\, l_m} = \mu_0 \frac{B_g a_g}{B_g l_g}$$

Or, the slope of the line OG, Fig.6.7(b),

$$tan\,\theta = \frac{B_d}{H_d} = \mu_0 \frac{l_m a_g}{l_g a_m} \qquad \text{...................................} (6.6)$$

The slope of this line is, therefore, decided solely by the dimensions of the air gap (l_g, a_g) and those of the magnet (l_m, a_m), and is independent of the shape of the demagnetisation curve. If the demagnetisation curve changes, may be due to magnetisation upto some different level, or due to some different quality of the steel, the point of intersection of any such new demagnetisation curve and this line OG always gives the operating point of the magnet. This is the reason this line is termed as the '*Load Line*' of the magnet. The quantity in the right hand side of Eqn.6.6 is termed as '*Unit Permeance*' of the system, and the load line as the '*Permeance Line*'. Eqn.6.6 also speaks another important fact. For given values of l_g, a_g and a_m, the slope $tan\,\theta$ is directly proportional to l_m. This means, shorter the magnet, less is the inclination of the load line, and less is the value of the flux. This effect is often referred to as *self-demagnetisation* of short magnets.

6.4.3 Minimum volume of permanent magnet: In permanent magnet design, economy in terms of cost and weight demands for minimum volume of the steel. Multiplying Eqn.(C) and Eqn.(D), we get,

$$(B_d.H_d).(l_m.a_m) = \frac{B_g.l_g}{\mu_0} B_g.a_g, \text{ or the volume of the magnet}$$

$$V = l_m.a_m = \frac{B_g^2.l_g.a_g}{\mu_0(B_d.H_d)} \qquad \text{.......................................} (6.7)$$

Table 6.1: Properties of permanent magnet materials

Group		Magnet material	B_R (in T)	H_C (in kA/m)	$(B_dH_d)max$ (in kJ/m^3	$\mu_{rrecoil}$
Metal	Iron alloys	1% Carbon steel	0.9	4	1.56	1.8 to 8
		6% Tungsten steel	1.05	5.2	2.62	
		35% Cobalt steel	0.9	20	7.80	
		Alnico group (Fe, Al, Ni, Co, Cu, Ti alloys)	0.75 to 1.35	40 to 120	12 to 84	
	Rare earth alloys	Samarium cobalt (SmCo) alloys: $SmCo_5$ and Sm_2Co_{17}	0.85 to 1.25	650 to 900	140 to 280	1.05 to 1.08
		Neodymium iron boron (NdFeB) alloy, $Nd_2Fe_{14}B$	1.2 to 1.3	750 to 980	278 to 318	
Ferrite (Ceramic)		Barium Ferrite, $BaFe_2O_3$; Strontium Ferrite, $SrFe_2O_3$	0.35 to 0.45	198 to 260	22 to 35	

Notes:
1) Old materials (of early 20th. Century) carbon steel, tungsten steel and cobalt steel are now obsolete.
2) Alnico and ferrite or ceramic materials were developed in 1930s and 1950s respectively, and still are in extensive use.
3) Rare earth group materials are latest developments (1970-1990).

For a given flux density required in an air gap of given dimensions, the volume of the magnet is, therefore, minimum when the product B_dH_d is maximum. In Fig.6.6(b), the B_dH_d product is plotted as curve B along H-axis against the various flux densities in the magnet along B-axis. The product exhibits a maximum value $(B_dH_d)_{max}$ at some flux density corresponding to the point M on the demagnetising curve. Hence, minimum magnet volume is achieved by shifting the load line from OG to OM. Air gap dimensions (l_g, a_g) being decided by the system requirement, the minimum magnet volume is achieved by suitable selection of the magnet dimensions from the following relations.

$$l_m = \frac{B_g . l_g}{\mu_0 H_d} \text{ (by Eqn.(C)), and } a_m = \frac{B_g . a_g}{B_d} \text{ (by Eqn.(D)).,}$$

where (H_d, B_d) are the co-ordinates of the point M. On the approximation that the demagnetisation curve is a rectangular hyperbola, which the former closely resembles to, the point M is the point of intersection of the line OL and the demagnetisation curve. The maximum value of the $B_d H_d$ energy product, $(B_d H_d)_{max}$, per unit volume of the magnet steel and magnetised upto saturation level, is the best criterion of the steel material for its suitability as permanent magnet. Properties of principal magnet materials are given in Table 6.1.

6.4.4 Stabilization of Permanent Magnets:

In Fig.6.8, ACB is the demagnetisation curve of a magnet steel obtained with soft iron pole shoe. Point C is the operating point of the magnet with normal air gap, so B_1 is the flux density in the magnet. Line OC is the load line for the magnet and the air gap. If the air gap is then bridged again by the pole shoe, the operating point will move to point D, not to point A. If the air gap is next restored, the operating point will back to point C. Thus the demagnetisation curve ACB transforms to curve DCB due to the insertion of the air gap. If the magnet, during its normal use, be subjected to some temporary disturbing field which has the effect of increasing the air gap, the operating point will be pulled down to some point like E, with a temporary load line OE. If the disturbing field is removed, the new operating point will be G, the point of intersection of the original load line OC and the recoil line EF. A temporary disturbing field thus caused a permanent reduction in the strength of the magnet (reduction in the flux density from B_1 to B_2 in the magnet, and so in the air gap too). Such a change obviously cannot be accepted. As a counter measure, '*stabilization*' is carried out by deliberately applying a fairly large amount of disturbing field which pulls down the operating point from C to somewhere H. On removing this stabilizing field, the operating point recoils to the new permanent point K on the original load line. If now the magnet be subjected to, say, the same temporary disturbing field, the operating point will move temporarily from K to L, and on removal of the disturbing field, return back to the point K. There is thus no change in the new permanent operating point K. It can be seen that JKHB is the final demagnetisation curve after stabilization. Stabilization, therefore, caused a reduction in the strength of the magnet (flux density in the magnet reduced from B_1 to B_3) but ensured its permanency against ordinary disturbing fields. Effect of stabilization is taken into account during design of permanent magnets.

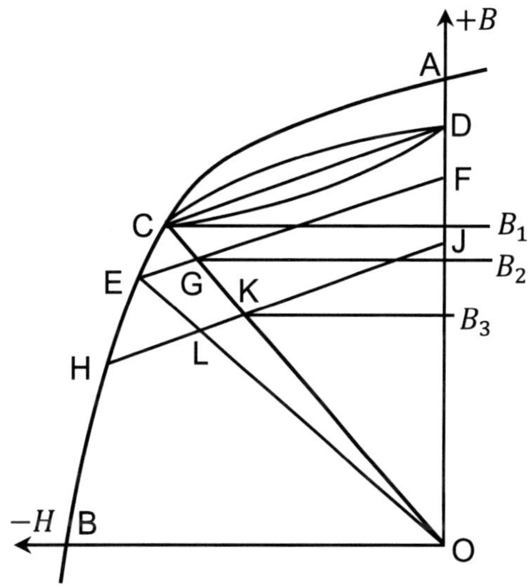

Fig.6.8 Stabilization of permanent magnet

An alternative view of stabilization follows. As the operating point of a magnet rises up the demagnetisation curve, better above the knee portion, the magnet becomes more and more immune to disturbing fields, so its stability increases. Now, the demagnetisation curve ACB of the magnet steel with pole shoe modifies to curve DCB with normal air gap, to curve FEB with the disturbing field, and finally to curve JHB after stabilization. In the process of stabilization, therefore, the operating point K has been favourably pulled up the demagnetisation curve JHB, well above the knee portion around point H, as compared to the original operating point C on the curve DCB. Hence, the stability of the magnet has been enhanced by stabilization.

Appendix

A1 Some Mathematical and Physical data used in the book:

Circumference / diameter ratio of circle, π	3.14159265........ $\cong$ 3.1416 or 3.142
Base of natural logarithm, e	2.7183
Permittivity of free space or primary electric constant, ε_0	8.854×10^{-12} farad.metre^{-1}
Permeability of free space or primary magnetic constant, μ_0	$4\pi.10^{-7}$ $= 1.257 \times 10^{-6}$ henry.metre^{-1}
Free-space velocity of light, c	2.998×10^{8} metre.second^{-1}
Mechanical Equivalent of Heat, J	4186 joule.kilocalorie^{-1}
Specific heat of water at 15^{0}C	4186 joule. kilogram^{-1}. Celsius degree^{-1}
Electronic charge, e	-1.601×10^{-19} coulomb
Electron rest mass, m	9.107×10^{-31} kilogram
Radius of electron, proton or nucleus	Of the order of 10^{-15} m
Radius of atom	Of the order of 10^{-10} m

A2 Multiples and Sub-multiples in Decimal System:

10^{12}	tera	T
10^{9}	giga	G
10^{6}	mega	M
10^{3}	kilo	k
10^{2}	hecto	h
10	deka	dk
10^{-1}	deci	d

10^{-2}	centi	c
10^{-3}	milli	m
10^{-6}	micro	μ
10^{-9}	nano	n
10^{-12}	pico	p
10^{-15}	femto	f
10^{-18}	atto	a

Lakh	$= 10^{5}$
Million	$= 10^{6} = 10$ Lakh
Crore	$= 10^{7} = 10$ Million
	= 100 Lakh
Billion	$= 10^{9} = 100$ Crore
	= 1,000 Million
	= 10,000 Lakh

A3 Notes on use of Names, Symbols and Prefixes of Units and Abbreviations:

(Based upon the approval of International Bureau of Weights and Measures)

1. Unit symbols should remain the same for singular and plural, e.g., 10 cm, *not* 10 cms, (because the latter stands for 10 centimetre-second). This restriction is, however, relaxed if the name of the unit is written in full, e.g., 10 centimetre and 10 centimetres both are correct.
2. In writing full name of a unit, the first letter should not be in upper case, e.g., 10 ohm, not 10 Ohm.
3. A unit symbol should be used only after some numerical value, not after any alphabet, e.g., 5 m, L metre.
4. The abbreviated forms AC and DC should be used only as adjectives, e.g., an AC circuit, a DC motor.
5. A hyphen should be used between the numerical value and the unit when the combination is used as adjective, e.g., a 100-W lamp, or a 100-watt lamp.
6. A slash (/) denotes division, e.g., kg/m^2
7. In writing unit symbol, use of multiple slashes should be avoided, e.g., m/s^2, *not* m/s/s. It is advisable to write unit symbols in the form $m.s^{-2}$
8. A unit symbol should be placed after the complete numerical value of the quantity, and should not end with a final full stop or dot, e.g., 10.345 cm, *not* 10.345 cm.
9. Full stops should be used in a multi-word abbreviation, e.g., e.m.f., p.d.
10. In writing the symbols for compound units, dots should preferably be used between the individual units. Such dots may, however, be dispensed with only if there is no risk of confusion. For example, the symbol for newton metre (unit of work) may be written both as N.m or Nm, but the symbol for metre newton (unit of torque) must be written as m.N only, not as mN which latter denotes milli newton.
11. Multiplying prefix should be used before any unit symbol *without* any space or dot in between, and the combination should be raised to some positive or negative power *without* using brackets, e.g., cm^2 means $(cm)^2 = 10^{-4}\ m^2$, whereas it is not allowed to write $c(m)^2$ to mean $10^{-2}\ m^2$
12. Multiple multiplying prefixes should be not be used when single equivalent prefixes are available. For example, pF should be used in place of $\mu\mu$F. Similarly, Mg, *not* kkg.
13. A multiplying prefix should preferably used in the numerator rather than the denominator, e.g., g/cm^3 should preferable be written as Mg/m^3.
14. Numbers of the form $n \times 10^{\pm p}$ should preferably be so expressed that p is some simple (integral) multiple of 3. For example, free space velocity of light 2.998×10^8 $m.s^{-1}$ should preferably be written as 0.2998×10^9 $m.s^{-1}$, or as 0.2998 $Gm.s^{-1}$. Accordingly, the multiplying prefixes centi, deci, deca and hecto, which do not satisfy this condition, should preferably be avoided.